To the fair
Love, Carolyn

DESIGNING WITH NATURAL PROCESSES

BUILDING AS IF YOU ARE PART OF NATURE

ACKNOWLEDGEMENTS

I acknowledge with gratitude the financial support of the National Science Foundation, the University of Illinois, National Academy of Science and Natural Process Design, Inc., the company I founded and own.

INTRODUCTION

My world was full of possibilities. I studied philosophy and found inspiration in "process philosophers" from Bergson to Whitehead who said that change is more important than constants. As I went on to study architecture at a progressive art school, I tried to replicate and incorporate the elegance of change and transformation that occur in nature into my building designs. I started with buildings that move and change like natural forms. *But someone told me that you really need to go the material level in your designs to build like nature.*

After becoming a registered architect teaching architecture at a university, I realized that architecture really did not teach—or design—with nature or the environment in mind. I wanted more; I wanted man-made environments to help, not fight, with nature. I studied evolution to conceptualize designs that incorporated processes from nature and mimicked natural forms and processes. I completed a master's degree in architecture in which I studied biology, ecology, and other nature-related science. Through these studies I realized that nature works, at its core, through the chemistry of its materials and their interaction with the environment. I had set on the path of incorporating the beauty, economy and inter-relatedness of natural engineering and systems into my work—and at its core that meant processes based in physics, chemistry and materials science.

Chemistry set ups

This book is about that passion in my life: *Natural Process Design.* It is not about designing with natural materials, or fitting in with nature, or about sustainability. It is about under- standing how nature builds in a beautiful, functional way and then applying that under- standing to the creation of buildings, airplanes, ports, bridges, coatings for roadways—the list is endless.

When examining a material that you may wish to build with, you need to find a process in nature that allows you to place that material strategically or form with it strategically, allowing for the possibilities of self-formation and repair, self-sensing, and recycling. For example, when designing a port for the Navy, I looked to ions in seawater for a model. Ions behave like particles, are capable of carrying an electric charge, and are found in an electrolyte medium which can transmit a charge; that is, they are found in seawater. If stimulated with metal electrodes which attracts charged ions, a structure is created out of these ions which flow to it through seawater. *It looks and acts similar to concrete.*

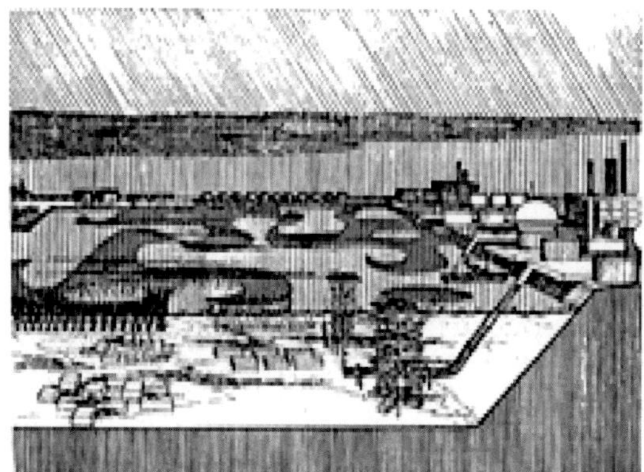

Drawings of my self forming port

Over time ions can be attracted again if needed to repair any damage to the structure because the ones in the structure can self-charge as they are "pressure electric" or piezoelectric (A piezoelectric effect describes the ability of materials to generate an electric charge in response to mechanical stress. The word comes from Greek words for squeeze, press, and push.) For self-sensing, these same ions in seawater can flow across a charged membrane to deliver a signal. For example, if you turn off the metal electrodes for a long time, ions from the structure will migrate back into the seawater, thus recycling themselves. This book contains other examples of designing with natural processes but I have always found this example to be the most fully expressive of the aspects of designing with natural processes and a simple way to tell a much larger story.

Designing with natural processes requires that elements of materials interact with the environment to function as part of nature instead of sitting apart from or defying it. This way, instead of creating a negative environmental impact, our impact becomes a symbiotic relationship of mutual benefit. As nature's processes and energies are much stronger than anything made by humans, the choice is really not ours to make. In time, nature will either take down what we make or evolve in harmony with it. As an inventive species, we have created

Photo of ocean, a port

massive sustainability problems for the future to deal with. Designing with natural processes offers an optimistic solution to some of these issues.

To design like nature does, I first needed to understand the chemistry responsible for natural forms and their dynamism. As an architect and creative person, I was motivated to emulate the aesthetic beauty of nature and I realized that to build the way nature builds I needed to understand environmental interactions both in form and in chemistry. Nature has developed self-sustaining processes that self-form, self-sense, self-repair, recycle, and improve its future environment. I have worked with this concept for over 40 years—creating solutions based in chemistry as well as designs inspired by and embracing natural systems. By employing natural processes in cooperation with nature I create and build with the recognition that *our human world is entirely within nature.* Over the years I have happily turned my back on the artificial, "removed" architectural approaches I find so ubiquitous. The results are a form of biomimicry I have come to call the *Natural Process Design*.

NPD requires a designer to make a close examination of the natural world we are all sustained by and from which we have emerged. Millions of years of evolution offer endless success stories to study and mimic. How will we emulate these survivors?

WHAT IS NATURAL PROCESS DESIGN?

Nature builds by

- using local, inexpensive, available, often recycled materials;
- self-ordering or growing by attributes shared between the material and environment;
- repairing itself;
- sensing and adapting to daily, seasonal, and yearly changes in the environment;
- easily disintegrating and recycling back into the material sink when usefulness is at an end;
- not harming the environment, but enhancing it or resolving deficiencies within it.

Leaves and trees are nature's way of regulating the atmosphere

This book has examples of each part of this paradigm and the overall approach too.

If you combine processes of the environment such as heating and cooling with proper- ties of materials such as melting and solidifying, you have the simplest example of an intelligent structure: an igloo. The ice and snow allowing for winter protection melts in the summer when it is no longer needed.

Another example is the way that corals build their reefs. Calcium carbonate ions with an electrical charge occur in seawater and the electrical charge from the coral attracts and then attaches them as part of the coral body. Thus, ions, the electrolytic medium of seawater, and a charge from the coral are all needed as interactive elements required to build coral. Algae has another symbiotic relationship with coral by making carbohydrates from solar energy and providing a food source for the coral. Could we build a seaport like we observe coral being built? We would need attributes of the environment (the electrolytic nature of seawater), a material (chargeable ions), and a process (an electric charge) to pull the ions to our coral-like concrete and co-create a seaport.

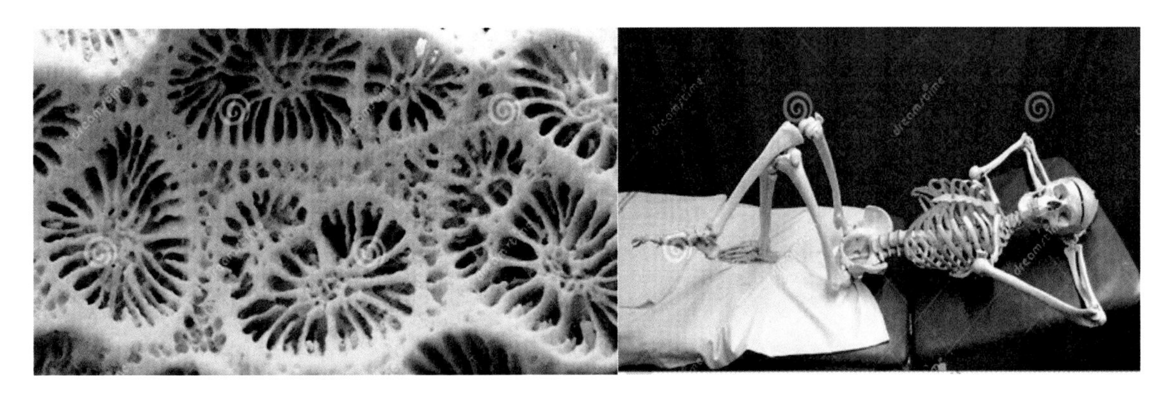

Corals and bones are examples of the similar expression of natural processes and materials

Human bones, nerves, and muscles use materials and processes similar to coral builders and creatures with shells because our bodies are based on the chemicals in salt water *(Figure 2)*. Chemicals in our seawater-like lymph can build bones, give off signals, and cause our muscles to respond. In fact, work is being done to create artificial bone by copying the way coral grows.

Coral reefs have symbiotic natural processes between the corals and algae- they cooperate

WHY DESIGN WITH NATURAL PROCESSES?

Natural Process Design is a biological mimicking system in which the beauty and intelligence of naturally integrated systems are studied and implemented for human use. Unlike other design systems, it integrates functions such as self-formation and repair, self-sensing, and recycling. This integration relies on using chemical behaviors as they are influenced by forces in nature and uses various natural processes. Often the forces of nature that cause what we label as destruction can be used as part of a remedy. You will notice that in many designs within the natural process paradigm, actions which can be very destructive such as corrosion, cracking, freezing, and thawing can be used as either a trigger or energy for a solution. Much like tai chi or judo, what could be destructive energy can be embraced and refocused into productive energy.

Imagine if we could design streets that repair their own cracks and subsequently extend the life of roads. What if there was a street surface that absorbed hazardous effluents of car and truck emissions, strengthening the concrete in the process and giving off oxygen? And what difference might it make if cargo ships and airplanes were able to use less fuel and had self-repairing properties simply by adjusting the chemistry of the materials used to build them? These things, and others, have been explored, tested, and proven possible using natural process design.

Nature's creative process uses cheap materials that are close at hand, sustains itself, self- senses; repairs damage; forms symbiotic relationships; recycles, and improves the environment for life that follows. Simply put, "mother nature" participates in the environment by using the chemicals and processes at hand to build the world.

The following chapters introduce examples of large projects. One of these project is fully integrated, meaning that the concept uses all the functions of the *Natural Process Design* at once; it follows a paradigm of design as if it is part of nature itself. The one that most embodies this is the Navy port project using the chemistry of seawater.

THE NEED FOR OXYGEN

The earth was formed in the same event with Mars, Venus, and the other planets from the big bang so they were very similar in composition but not size. Venus is very hot and Mars is cold and they cycle in temperature widely in one day. The earth has evolved differently to have a stable moderate temperature in a small range with a breathable atmosphere that can support animal life. How did this happen?

Views of earth from space

Our original atmosphere of gases, from the center of the earth and emitted from volcanoes, was toxic, full of methane, sulfur and CO_2. It was incredibly hot with very large heat fluctuations. Gravity held in the atmosphere and eventually plant life was formed from bacteria. Plant life transformed the atmosphere to an oxygenated one that animals could breathe. This atmosphere insulated the earth and stabilized the climate to a small range of steady temperatures that were cool and warm enough to support animal life.

An understanding of the atmosphere and earth's environment being hospitable to life helped me understand why our current fossil fuel economy is so destructive to the earth's atmosphere and environment. Due to plant life, earth went from an atmosphere of methane and lots of CO_2 released from volcanoes and the interior of the earth with large temperature swings daily to the one that exists today which has oxygen and stable temperatures in a small range. We are in the process of driving the atmosphere back to a carbon dioxide- and methane-rich one which, of course, is not good for oxygen-breathing humans and animals. Of particular concern is the die-off of phytoplankton in the Antarctic ocean. These organisms that produce about 50% of the world's oxygen and sequesters 50% of the world's CO_2 are being killed by excess sunlight under the ozone hole without an ice sheet over them. The ozone hole is as serious a problem as the build-up of greenhouse gases from burning fossil fuels which caused the warming that melted the ice cover.

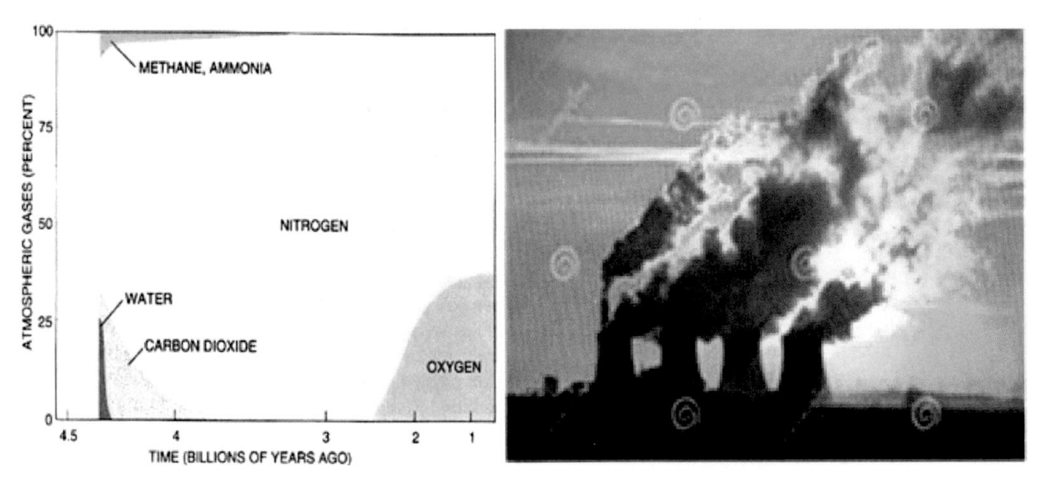

Left is a drawing of the earths atmospheric history over time, right is a photo of a power plant releasing carbon dioxide which moves the atmosphere now towards a reducing one

Volcanoes releasing gases from beneath the earth's crust; these gases are sulphur dioxide, carbon dioxide and other reducing gases such as methane and ammonia

Fossil fuel plants burning fuel and releasing carbon dioxide, a greenhouse gas that causes global warming

Essentially what we have achieved with burning massive amounts of fossil fuels is a reversal of atmospheric direction so that we have less oxygen to breathe and too much CO_2. A way of prioritizing responses to this climate crisis is to use the approach in natural process design. I always seek to find the basic, most important factors in a situation and those are not water or soil quality, but the more generative and basic factor: the gaseous atmosphere. This factor greatly affects all others. We need to be mindful of various aspects of a problem but it is important to prioritize otherwise people can favor easy solutions and use their energy, time, and resources without having much impact on preserving a livable environment.

The atmosphere insulates the earth and stabilizes the climate. A small range of steady temperatures are both cool enough and warm enough to support animal life. The atmosphere has oxygen which supports animal life but as we drive it back to one rich in CO_2 with less oxygen and a disproportionate amount of greenhouse gases, we are in danger of overheating; working against our own survival *(Figure 3)*.

OUR IMPACT ON THE EARTH'S ATMOSPHERE

- Can effluents from fossil fuel driven building and vehicle production be used to reduce the damage and help solve the environment's problems?
- Can the use of less concrete by making it self-repairing reduce CO_2 production given off during its creation?
- Can we prevent the pollution of ground water with heavy metals from fly ash, a product of burning coal?

Designs using natural processes can address the needs represented by each of these questions. A NASA computer model provides a portrait of carbon showing the amount of CO_2 produced by the earth and it is clear that the growing seasons of spring and summer in the Northern Hemisphere take up CO_2 while the volume of CO_2 in fall and winter rises considerably. Modeling in harmony with this paradigm, I am designing materials that take up the effluents of a fossil fuel economy and thereby, improve air quality. Materials designed with natural processes can absorb harmful byproducts of an oil-based economy, especially greenhouse gases produced by burning fossil fuels.

MY SOLUTIONS

- I have invented ways of taking up excess CO_2 gases while strengthening concrete.

- I have developed fly ash (a product of burning coal) panels that sequester the heavy metals in ash from coal-fired furnaces that otherwise would end up in our air or leach into our water supply.
- I invented a house paint or coating that can take up CO_2 and give off oxygen, thus improving the atmosphere for all living things

Self-repairing concrete is a perfect target for the work of designing with natural processes because cement curing gives off approximately 8% of the world's CO_2 greenhouse gases. Another application of natural processes might be in self-repairing airplane wings that can be 30% lighter and thereby save fuel. Cars, and planes and ships are all heavy air polluters. Can we make them lighter and therefore less fuel consumptive? Airplanes are especially worrisome because they put the pollutants high in the atmosphere and also affect the ozone layer.

Fossil fuels—coal, gas, and oil—are the basis of much of the built environment as well as energy sources. Polymers/plastics are made from oil. Cement is made by burning fossil fuels at high temperatures and the process releases CO_2 from limestone. These two examples are but a few production methods that degrade the environment. Therefore, decisions on the built environment can be a major part of the solution to the problem of environmental degradation. Taking up effluents of a fossil fuel economy is part of my creative work as I seek to design with the "mind of nature."

As an alternative to focusing on the next generation of fuel sources, I suggest designing buildings, infrastructure and vehicles to use as little fuel as possible, employing the structure and vehicle materials to:

- take up greenhouse gases in building materials;
- catalytically convert the gas effluent flowing out of cars and the heating, ventilation, and air conditioning systems (HVAC) of buildings to environmentally desirable gases such as oxygen while strengthening building materials;
- develop an integrated home appliance which can take our waste and CO_2 and give us water, oxygen, and energy for heating or cooling; and develop cooling

materials that do not give off ozone-depleting chlorofluorocarbon (CFC) but absorbs those chemicals instead. Chlorofluorocarbon is an organic compound that contains only carbon, chlorine, and fluorine. It is produced as volatile derivative of methane, ethane, and propane, also commonly known by the DuPont brand name Freon™.

WHAT WILL MEASURE SUCCESS?

The ideas presented in this book involve design with natural processes in an effort to take up the effluent of a fossil fuel economy and through this solve some of the problems of an overheated planet. They are not merely "good ideas" but have been scaled up beyond lab experimental proof-of-concept to commercial-level projects with the support of involved businesses and government agencies. The next step is to get them into wide use and application and to keep inventing more solutions.

CHAPTER 1:

GENERAL CONCEPTS AND PROJECT OVERVIEWS

One of the concepts I have designed, based on the *Natural Process Design* paradigm, integrates all functional aspects into one system, using one chemistry: my ocean port work. This is the best and most integrated example of using a biological system as the model from which to design a building system. It relies on the environment of available chemicals to integrate man-made and nature-made materials. In this ocean system, the body's liquid system of blood and lymph is the model.

The fact that the seawater is basically the same as human lymph allows us to do many animal-like functions when building in seawater such as forming bone-like structures with sensing, repairing, and motion functions. The important commonalities of seawater and lymph are that they contain many minerals and electrically-charged molecules called ions and can carry an electrical charge; i.e. they are electrolytes. (An electrolyte is a substance that produces an electrically conducting solution when dissolved in water. Electrolytes carry a charge and are essential for life. All higher forms of life need electrolytes to survive.) These ions and minerals can be "inspired" by electricity to form a structure.

Drawings of the self forming port

The ions can lose their charge when going across a charged membrane (such as occurs in human nerves) and the membrane can give off a signal. Like human bones these ions can be attracted by the electricity of the bone (or concrete-like structure) to repair a broken portion.

In the self-building ocean port project, the chemistry of seawater supports all of the functions required. The attempt to *emulate* this model of nature is a project of making in-situ building materials in the ocean from the chemicals in seawater. The electrolytic seawater can carry an electrical charge and move calcium ions onto the charged structure. This forms a coral or bone-like material.

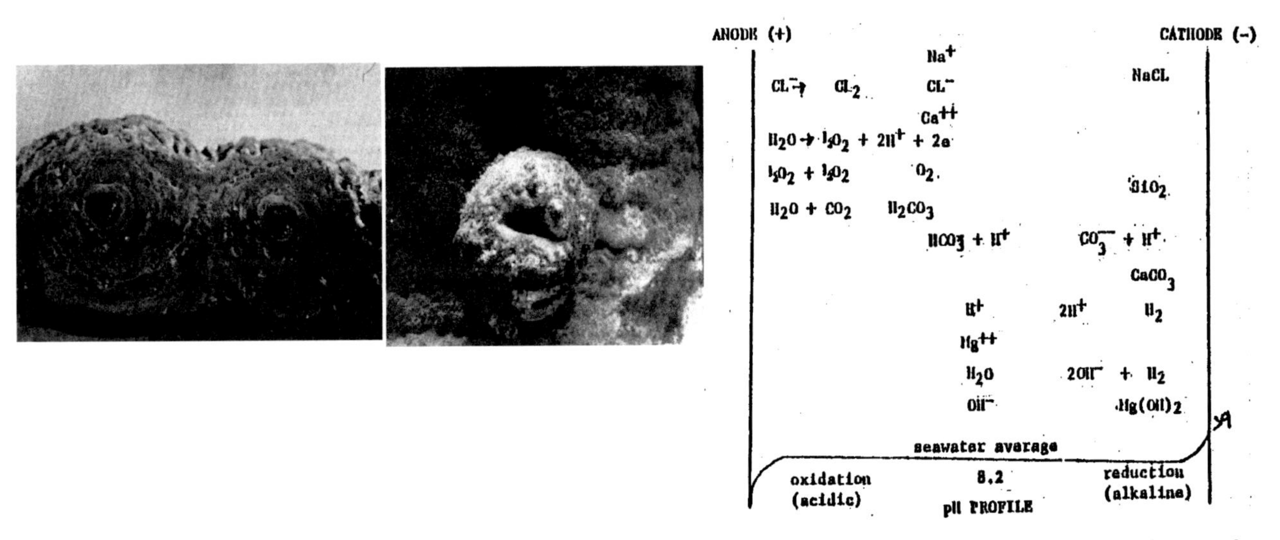

Left are photos of accretions made from the chemicals is seawater Right shows the chemicals in seawater

Because humans evolved from seawater, our lymph is like seawater in chemical composition. Therefore, I looked to the animal body for inspiration. The bone-like structure of the ocean port repairs itself by the addition of more stress which pulls out more calcium from the ocean, just like human ribs repair themselves by using the calcium in the

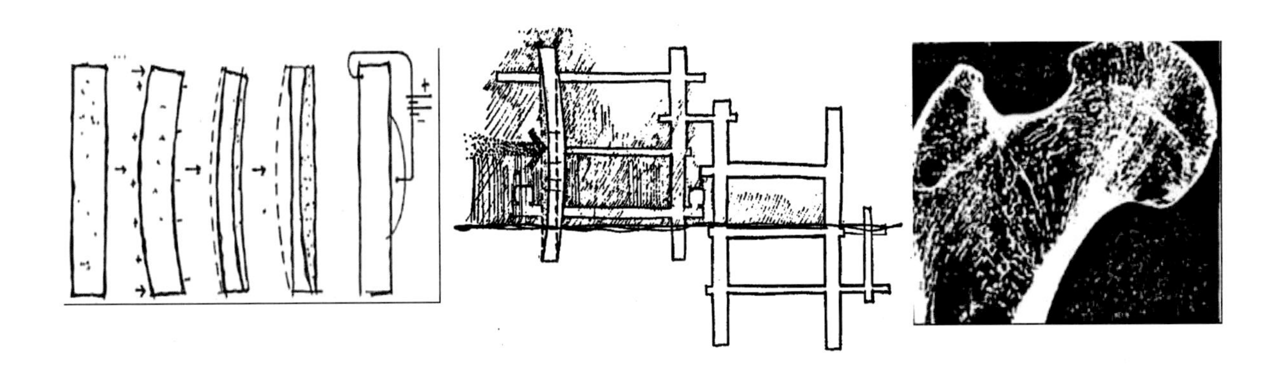

Left two are drawings of the self repairing of bones or columns, Right is a bone

lymphatic system. The port's "nerves" can sense charged ions going over a membrane and can move parts like muscles move by using different chemical concentrations in water to make an osmotic pump.

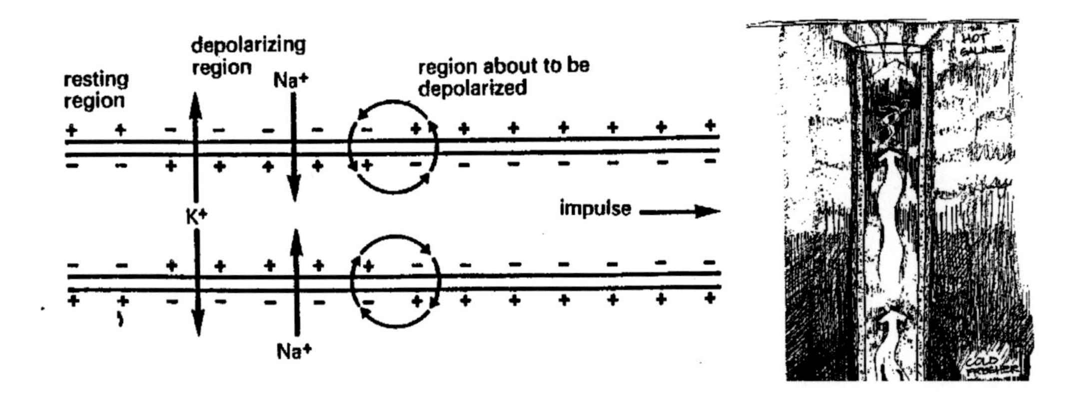

Left is a drawing for the chemistry of nerve communication, Right is a pump using sewater

This project emulates the complete biological paradigm over time of self- growth and positioning, adaptation and repair, and recycling. Calcium carbonate in seawater is electrochemically deposited onto a framework following the work of Wolf Hilbertz. This is accomplished by charging the seawater (the environment) with electricity which attracts the charged ions (calcium carbonate) to a framework; a process called accretion. Pier columns constructed of this material can adapt to changes in load pressure because the electricity generated under pressure (piezoelectricity) attract more charged ions from seawater as the pressure increased, thus be self-repairing. The idea for sensing and repair was to mimic nerve-initiating fibers which strip off electrical charges as ions flow through a nerve membrane. Employing various differences in the chemicals in seawater make motion possible. When the pier is no longer in use, the absence of pressure allows the charged ions in the structure to lose their electrical attraction and dissolve back into the seawater, thus be recycled.

PROJECTS WHICH ADDRESS PARTS OF THE PARADIGM: FLY ASH

I have developed several materials and technologies that address the issue of how to build with the natural processes of nature and concurrently take up effluents of our fossil fuel/ chlorofluorocarbon (CFC) economy (examples: CO_2 and coal ash) and prepare for a new non-fossil fuel/non-chlorofluorocarbon economy. Mother nature always wins so building with *Natural Process Design* is the only viable long-term path forward.

In individual natural process building projects which take up effluents from fossil fuels, the chemical bases are variable because many different effluents are involved.

Left are the fly ash sphere, next one of my panels fabricated, and right a brick sample.

Fly ash is the waste product of burning coal, contains many heavy metals, and is largely carbon. I developed a way of converting fly ash into building panels and insulation of nearly all ash so that these metals are sequestered and do not leach out to pollute the environment. Using nearly 100% fly ash by cooking it into a solid has another use by allowing you to cook it at a lower temperature, using less energy and fewer chemicals. Ash is usually put into cement as an additive, but only a certain percentage of the cement panels can be ash else excess dilution may occur.

Left, a soil compactor used on a fly ash panel at a company, right, another ash panel.

PROJECTS WHICH ADDRESS PARTS OF THE PARADIGM: SELF REPAIRING CONCRETE

Making cement produces 8% of the world's CO_2. Cement makes into concretes. Concrete is a very common material but is porous and reacts highly to intrusion from the environment. All of our solutions for preserving concrete address making it less porous or preventing damage such as corrosion and cracking as the environment intrudes. What if we strengthened concrete and closed off the pores so it could not be penetrated by chemicals and water? CO_2 reacts with the basic cement component–lime–to make the "glue" in cement and this can occur in the concrete itself without outside energy chemical reactions. This will work to revive its strength, extend its durability and lifetime while taking up carbon dioxide. I developed this self-repairing concrete where less cement needs to be replaced. Using less concrete because we make it more durable is an important place to intercede on behalf of the environment. We made four full-sized bridges of this self-repairing concrete and tested them.

Left are self-repairing concrete bridges, right is the chemical released from the bridge when it was cracked by pushing up from below.

Left, fibers with self repair chemicals are thrown into the cement mixer at the site, right they survive after mixing in the concrete.

PROJECTS WHICH ADDRESS PARTS OF THE PARADIGM: SELF-REPAIRING POLYMERS

Our polymer composites made of graphite from oil are self-repairing. In a project completed for the U.S. Air Force, it is expected that the composite portion of the airplane fuselage or wing can be 30% lighter with self-repair. The addition of self-repair allows the composite to have fewer layers than normally used to prevent delamination. This would save large amounts of fuel and CO_2 emissions; an airplane trip uses three times the gas than a car trip of the same distance.

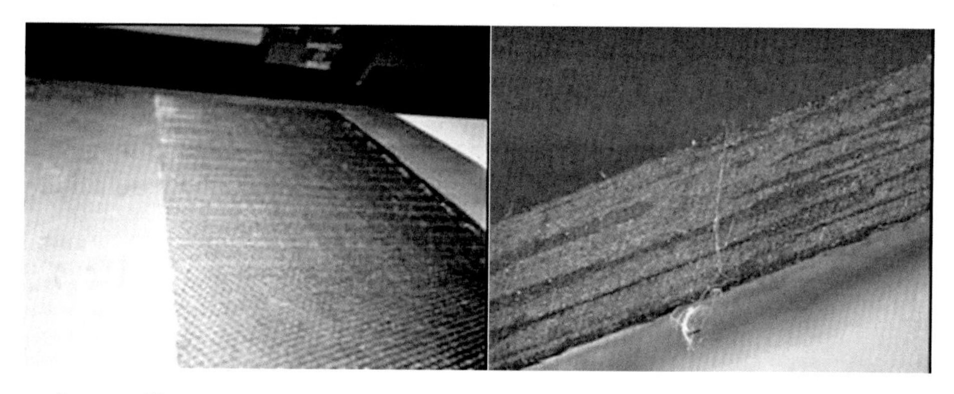

Left, part of our self-repairing airplane wing. Right, a slice through the graphite composite. Dark areas of repair chemical can be seen.

New planes and ships are being made from polymers in order to be lighter because of their large part in global warming. Most cargo is sent by ships and vapor trails of planes release CO_2 and chlorofluorocarbons, a double whammy which causes both global warming and damage to the ozone layer. The polymer composites currently in use delaminate and break easily so manufacturers don't trust them and make ship and plane components thicker and heavier to compensate, reducing the advantages of the lightweight polymers. Self-repair eliminates these concerns and is estimated to reduce the weight by 30%.

Even walls that can resist blasts can self repair so that the repairs have gone from a one inch size delamination to a damage delamination of four feet.

Left is self-repairing wall panel after blasts. Next is the blast set up. Right is a blast to test projectile speed.

PROJECTS WHICH ADDRESS PARTS OF THE PARADIGM: SELF-REPAIRING POLYMERS FOR PRESSURIZED SYSTEMS

One of the most useful applications of self-repair is in pressurized composite structures. For NASA, a self-repairing wall for a pressurized habitat was developed that could withstand penetration and yet hold pressure when exposed to a vacuum. Polymer walls can be self-repairing. In the research, I moved from self-repairing delamination damages of 4-6 inches in diameter to damages of 12-14 inches in diameter, as well as up to 48-inch diameters that survive blasts. Other uses of self-repair in pressurized systems were tires for bikes and cars.

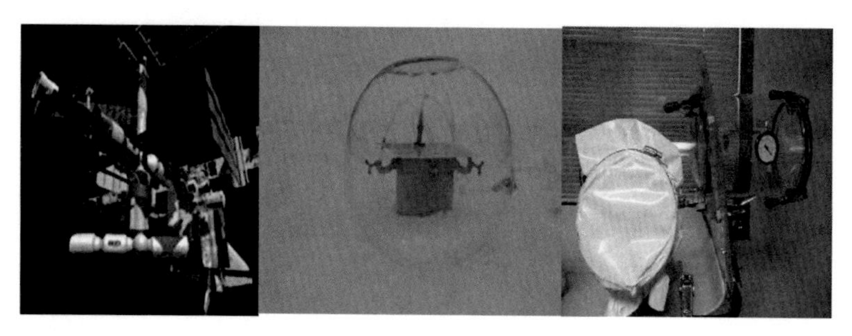

Left, the International Space Station has capacity for materials experiments both inside and out in space, Right two, proposed space experiments for self-repairing pressurized infrastructure for testing on space station.

Left, puncturing a truck tire, right, inside the repaired tire sample, pressure was applied and measured and hoop strength calculated

PROJECTS WHICH ADDRESS PARTS OF THE PARADIGM: SELF-SENSING

These samples of self-repairing materials can self-sense to assess the state of damage and repair. Self-sensing is the twin to self-repair. Damage can go undetected in polymers as well as in concrete and catastrophic failure can result. Providing knowledge of the damage and the repair accomplished allows manufacturers and users to be more confident in their use of lighter weight materials. Self-repairing and self-sensing materials could help materials and trucks and pressurized pipes for oil and water or other fluids survive longer and reduce the use of oil.

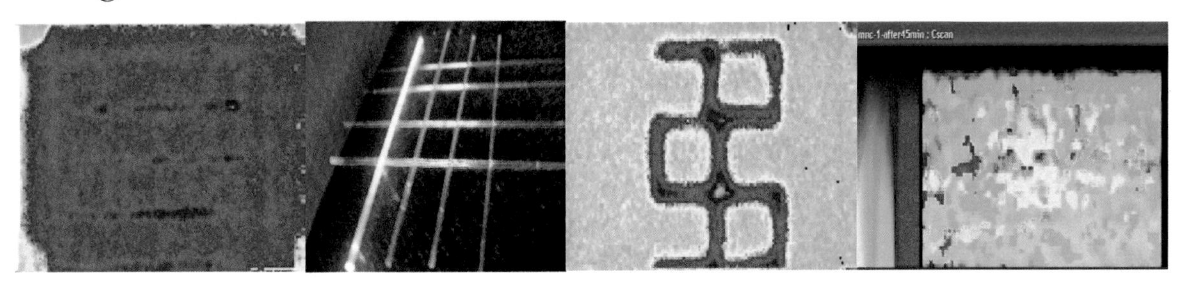

Left, a view of eddy currents that can emanate from metal coated fibers to locate damage. Center, a photo of fibers placed at right angles to each other to indicate, by sensing of fiber breakage, where damage is located in 3 dimensions. Right, metal coated fiber grid sensed by eddy current scans at 2 KHz. Far right, pictures of an ultrasound test.

PROJECTS WHICH ADDRESS PARTS OF THE PARADIGM: RECYCLING BRIDGE GIRDERS

Pre-stressed girders are the most difficult construction component for demolition because they have an embodied energy of high tension metal rods that can break and blow the whole girder and surrounding structures apart when cut. In an explosion they tend to destroy all the material around them as the pre-stressed tension in the steel tendons is released. Most bridges and many buildings use these types of girders. The natural process design solution has built-in ways of reducing the tension before demolition; a way to demolish materials with minimum damage to the surroundings. I developed a dismantling technique which allows the metal rods be taken out while the rest of the bridge remains intact.

Photo of the girders being fabricated.

PROJECTS WHICH ADDRESS PARTS OF THE PARADIGM: COATING THAT ABSORBS CO2 AND STRENGTHENS THE CONCRETE AND SO IMPROVES THE ENVIRONMENT FOR ALL SPECIES

Much of the CO_2 released during decomposition of calcium carbonate (limestone) to calcium oxide (lime) and carbon dioxide can be taken up by the concrete over time. I developed a paint that utilizes this ability and speeds it up. It can be used on highways or buildings. It takes up carbon dioxide in roads at the source (automobile tailpipes) and the reaction strengthens and repairs the concrete and extends its lifetime while improving the environment.

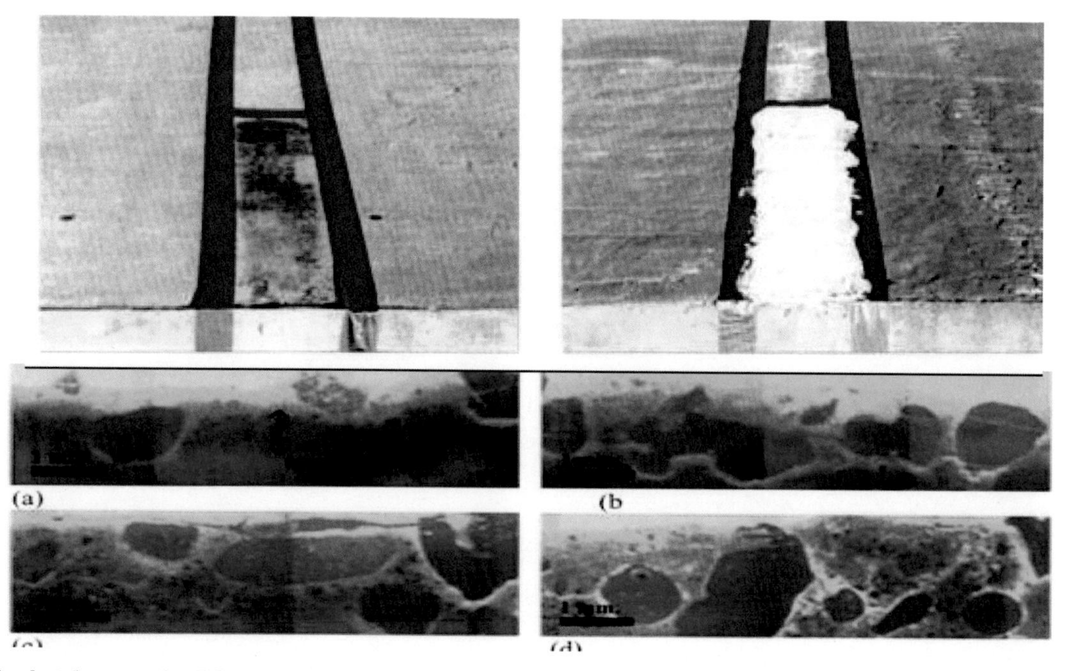

Top are the bridges with CO2 sequestering coatings, below a slice of the concrete showing CO2 penetrated.

PROJECTS WHICH ADDRESS PARTS OF THE PARADIGM: POLYMER/CERAMIC COMPOSITE BASED ON RULES OF BONE GROWTH

Nature self-forms materials by a process which involves uniting a property of the environment with a property of the material. Following the analogy of bone growth, I developed a cement and polymer material that forms like bone: the polymer forms a template for the cement to follow. It has the strength advantages of hard cement and pliable aspects of plastic, can be self-repairing, and is extremely durable.

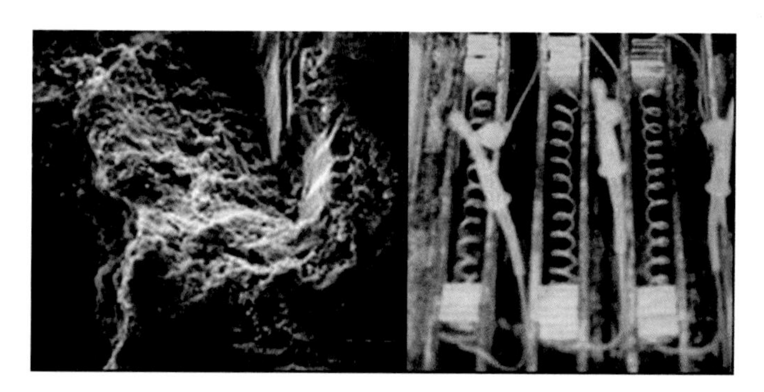

Left, a microscope photo of the cement/polymer material, right, the chemical delivery system for making this material in a mold.

There is an elegance found in the development of polymer (like bone development) through its creation. First the polymer is formed, which gives off water and hydrates the cement, then the cement and polymer give off heat to set up the polymer. Unlike usual cement creation, there is no outside energy needed for the reactions and no effluent released because it is taken up as part of the second reaction, forming cement. Because of these unusual properties, this could find medical applications as a bone replacement. Bones developing over time from a sequence of events were used as the model.

ADVANTAGES OF THE NATURAL PROCESS DESIGN SYSTEMS

- Ocean ports can take up CO_2 and give off oxygen.
- Waste materials from coal-fired furnaces such as the particulate matter fly ash are reused instead of being released into the air and heavy metals are prevented from entering ground water if ash is disposed of in other ways
- Self-repairing concrete lasts longer than conventional concrete. By extending its

lifetime we can reduce the 8-10% of the world's CO_2 created by cement manufacturing.

- Self-repairing polymer composites for airplane wings can reduce the weight of portions of a plane by up to 30%. This saves huge amounts of fuel and reduces air pollutants released into the high atmosphere
- Self-sensing can assure airplane companies that *Natural Process Design* works.
- Self-repairing polymer composite walls last longer and provide a different outlet for oil than burning it.
- Self-repairing tires allow petroleum-based rubber to last longer, which provides a better outlet for oil than burning and prevents tires from entering landfills.
- Self-repairing high-pressure composite pipelines mean we will have fewer pipeline leaks and less of the environmental damage these leaks cause.
- With the use of self-sensing, many items can be reused if there is clear evidence that repair has occurred.
- With a coating that absorbs CO_2 and other gases from the air and converts them into oxygen and water, buildings can be catalysts for improving air quality while strengthening concrete in the process.
- Demolishing concrete girders without ruining the rest of the bridges saves lives and extraneous concrete girder creation.
- A ceramic and polymer composite based on bone biomimicry can create a unique material with properties only found by learning from nature. Ceramic, in general, is strong but brittle and polymer is ductile and tough but weak in compression (as in bone). Our new material has the best of both worlds; it will have a longer life than either because it can resist various kinds of damage.

NATURAL PROCESS DESIGN BENEFITS EVERYONE

Application of biomimetic rules to take up effluents of a fossil fuel economy in funded research projects as described were made into full-scale projects with private or government support. These new ways of building combine materials and environmental properties to increase both monetary and environmental savings.

We will see this throughout the next chapter in adaptive ocean ports made from chemicals in seawater. Based on biomimicry, they take up CO_2 and give off oxygen. Self-repairing, self-sensing concrete saves cement and reduces the world's greenhouse gases. Self-repairing polymer composites use less petrochemical oils and the materials using fly ash sequester heavy radioactive metals. Creation using the paradigm of *Natural Process Design* profits government, business and the viability of Earth.

Reference cited

1. Wolf Hilbertz, 1979 Electrodeposition of Minerals in Sea Water: IEEE Journal on Oceanic Engineering, Vol. OE-4, No. 3, pp. 94–113,

CHAPTER 2:

MAKING OCEAN PORTS USING THE CHEMISTRY OF SEAWATER

Ocean port design in the main sense has been based on technology devised for use on dry land and for less climatically extreme environments. The ocean's volumetric movement of water (in the form of waves and currents) is viewed as a major problem for these structures. Seawater which contains vast quantities of minerals also can dissolve most substances. It corrodes most structures placed in it by covering them with encrustations of minerals and organic matter. The ocean and climatic environment are opportunities for new ways of designing with natural processes. My paradigm emphasizes building with the physical, chemical, biological, and climatic processes of the ocean and bordering regions. Building using natural processes is to build using only the most common and abundant resources, and organizing these by means of technologies which are indigenous to the environment. Natural process design also emphasizes self-regulation by a process integral to the system's material and organizing systems. The self-regulating process entails a way of sensing a need for change, re-organizing, re-collecting material, and re-distributing it according to the new organization. Thus, there is material, a way of distributing the material, some organizing principles, a way of sensing, and a means of reclaiming material.

This design philosophy draws from the "experience" of evolution and integrates those principles into technology. It is an examination of properties of entire systems and elements, their effects on each other and their performances based on interactions of a complex system. Ultimately, the capacity to organize and reorganize material in a complex way depends upon the relationships among elements, not upon the number of elements added. The advantages for building ports in the way described later in the chapter are potentially enormous. Savings come in the form of energy normally used for prefabrication, the imported material required, installation time, repair time and costs of replacement and obsolescence.

HOW DID THIS HAPPEN?

It all started when I was a new professor at Texas A&M University and needed to obtain research funding. I recognized who Athelstan Spilhaus was–a former dean of University of Minnesota Institute of Technology–so I interviewed him about needs of the sea grant program. I knew that there were possibilities, so I applied and received a grant. My work had just begun!

How could a port in such a sensitive environmental area as the Gulf of Mexico with its rare and endangered species be built with less damage? The models of biological building systems in the ocean were mangrove swamps and coral reefs.

Concept of port that can be built from seawater.

WHY THE OCEAN PORT PROJECT IS IMPORTANT

Not only are endangered species important, but the survival of the earth as a habitable place depends upon what we do with the oceans. As a matter of fact, Antarctic phytoplankton are said to be responsible for 50-60% of the carbon dioxide sequestered worldwide and the same amount of our oxygen generated. However, they are dying due to melting of protective sea ice through global warming and solar radiation from the ozone hole. The ocean port project builds in a way that does not introduce new harmful materials into the ecosystem and aims to build a haven for ocean species akin to reefs.

I asked myself how we could build as if we are part of nature and do no harm. I decided to try to integrate all the functioning of a coral reef into the port and I looked to the animal body as my guide, due to the lack of available literature on coral. I knew that the human body's blood and lymph have nearly the same chemical composition as seawater and that the body has all the functions of the paradigm. Here I decided the body's chemistry system would be my model for port building based on natural processes.

Seawater and lymph contain minerals and electrically-charged molecules called ions that can carry an electrical charge: electrolytes. That means that these ions and minerals can be moved to form a structure by electricity. The ions can lose their charge when going across a membrane (like how nerves function) and give off a signal. The electrolytic sea water can carry an electrical charge and move calcium ions onto the charged structure to form a coral- or bone-like material. Most importantly, these ions can be attracted by electricity of the bone- or concrete-like structure and repair a broken portion.

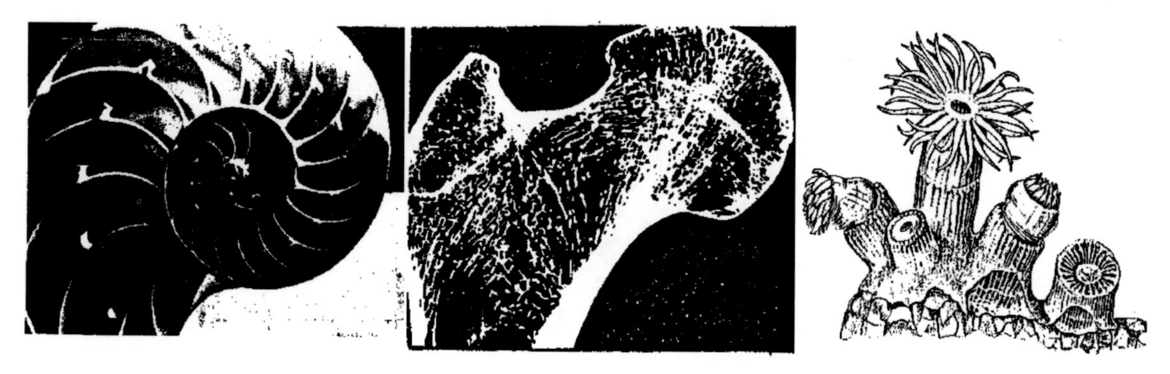

A seashell, bone and coral are all made up of the same materials and by the same processes.

The similarities between human lymph and seawater led me to other observations. Bone-like structures in the ocean self-repair when experiencing stress which extracts calcium from the ocean, just as your ribs might repair themselves by using the calcium in lymph. The port's "nerves" sense using charged ions going over a membrane and can move parts (like our muscles do) by using different chemical concentrations in water to make an osmotic pump.

This project emulates the complete biological paradigm: self-growth and positioning, adaptation and repair, and reuse and recycling. Self-growth is attained when calcium carbonate in seawater is electrochemically deposited onto a framework (following the work of Wolf Hilbertz) by charging the seawater with electricity which attracts the charged ions (calcium carbonate) to a framework (accretion) using solar panels as the energy source. Changes in load pressure in the pier columns are also addressed as the electricity generated under pressure (piezoelectricity) attracts more charged ions from seawater. Sensing and repair mimics nerve initiating fibers which strip off electrical charges as ions flow through. Self-repairing occurs where there is rubbing and subsequently piezoelectric signal, thereby attracting the charged ions in the electrolytic solution and then depositing them onto the broken area. A sensing example includes charged ions of seawater going across a charged membrane to give off a charge signal for sensing or a way of sensing exterior environmental changes and communicating that information and adapting to changes in the environment daily, seasonally, and yearly. Finally, recycling is achieved by the material disintegrating and returning into the sink of materials from which they came. When the pier is no longer in use, the absence of pressure allows the charged ions in the structure to lose their electrical attraction and dissolve back into the seawater; just like if one turns off the electrical stimulation in bones due to the lack of exercise, calcium gets washed out.

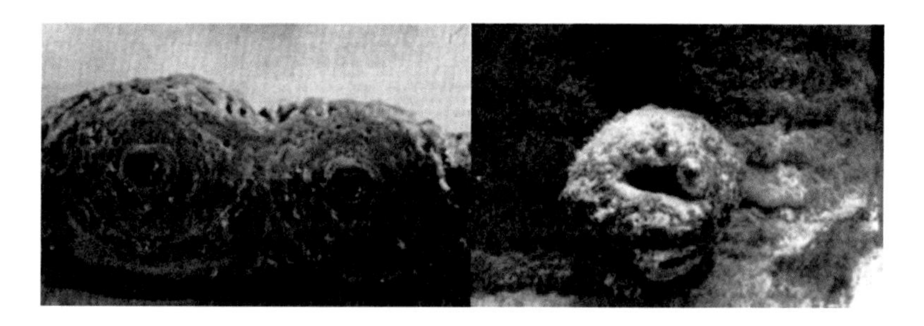

Left. accretion on two ½ inch hardware cloth cathodes. Right, detail of accretion near St. Croix.

This way of building can also save vast amounts of fossil fuels. Ordinarily building a port would use large amounts of steel and concrete and produce huge volumes of CO_2. The ocean forces are formidable so there is usually a lot of anchoring, reinforcing, and pumping required. In my port concept, natural processes design is applied to the structure by:

- using available materials which are as cheap and nearby as possible to eliminate purchase and transport costs;
- preferably using indigenous or recycled materials
- putting into organization or constructed or assembled by natural process means, often with an indigenous energy source;
- self-organizing on the chemistry level by some property of the environment and material such as ice freezing and melting as temperature changes, or bones accreting

by charged ions moving in and electrolytic solution in saline liquid;

- improving the overall environment in the process (ex: carbon dioxide taken up and oxygen given off);
- having self-sensing, self-repair, and regulation through a process integral to the system's material and/or the organizing system.

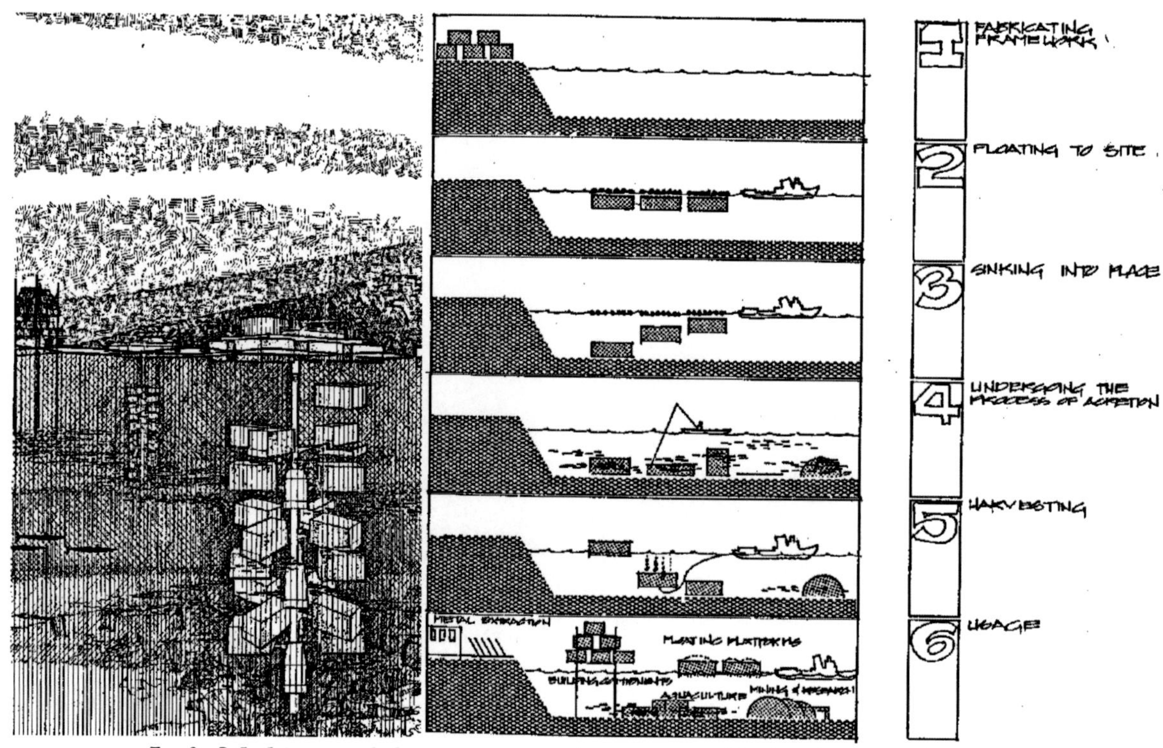

Left, Making modular components in seawater. Right, the overall process.

In the ocean port project one chemical system can support all the various functions required. We will see in the next projects that take up effluents from a fossil fuel economy while building as part of nature the chemical bases vary due to the different effluents involved.

ORIGIN OF OCEAN PORT PROJECT

This project was done at Texas A&M under at National Oceanic and Atmospheric Administration Sea Grant Program and later under an Office of Naval Research grant. The goal was to construct an ocean port in a location as environmentally sensitive as the Gulf of Mexico.

Seawater contents

Beside hydrogen (10.82%) and oxygen (85.84%) seawater contains the following abundant resources by mass and 51 other minerals and elements.

Chloride..... 1.94% Sodium 1.08%

Magnesium .. 0.1292% Sulphur 0.091%

Calcium ..... 0.04% Potassium .. .0.04%

Bromide..... 0.0067% Carbon 0.0028%

The means to manipulate clay, sand, rocks, ice, and organic materials are to dissolve, evaporate, solidify, liquify, plastify, elastify, associate, and dissociate (as ions), decompose, and compose. Some of the most indigenous technologies or means to distribute and gather materials are: 1) flow of fluids 2) electrical current and ionic movement in seawater 3) gas

flow and 4) gravity. The means to change a material's properties are chemical reaction, electrical charge, and ionic charge.

Sensors most readily applicable to the ocean environment are: chemical, thermometer, pressure, wave, photosensor, microphone and electrical.

CONCEPT OF PORT STRUCTURE

Ocean port design using natural process design has the possibility of employing reversible natural processes as building construction technologies for ports which are economical, energy-cheap and ecologically sound. Such processes are especially applicable to port design where time requirements are fluid and space needs are stringent. Thus, they may require speedy use of indigenous materials in a variety of available settings. In a remote location the importation of materials, labor, energy, and supplies are costly.

The use of solar energy to give electricity to electrochemically deposit calcium carbonate from seawater onto a structural framework follows the work of Wolf Hilbertz. 1 This is accomplished by charging the seawater (the environment) with electricity which attracts the charged ions (calcium carbonate) to a framework, imitating bone and mineral accretions like seashells. The ability of seawater to dissolve and corrode many substances in it and cover them with encrustations of minerals and organic matter are harnessed. These natural accretions have been found to have the bearing strength of concrete. Corals and sea mollusks make their shells by accreting minerals through use of an electric potential.

Metals in seawater are subject to such electrolytic processes and ions can be precipitated onto surfaces. A metal armature is hooked to electrodes with a charge and submerged in seawater and the electrical current produces an electrolytic deposition of minerals onto the armature framework.

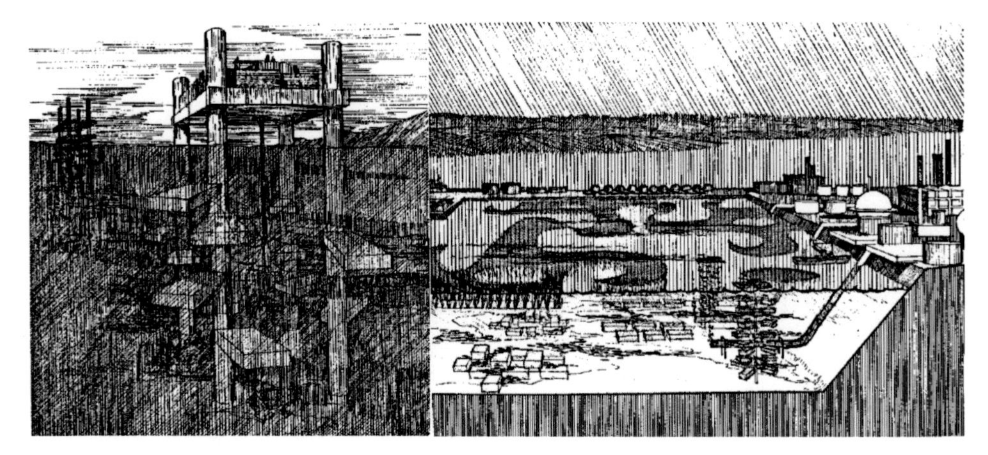

Left concept of port that can accrete portions that can be moved up, down and sideways, right concept of the port processes.

"The cathode gives off hydrogen; the anode, oxygen. A power supply from a battery charge of about 4.8U at 200mA to hardware cloth, anode of carbon about 10cm x 2cm in an aquarium 3 x 1½ x 2 feet for a period of 500 hours produces an accretion of 10mm.1 Using anodes of lead, carbon, or iron, the electrochemical reactions follow. The material is accreted due to concentration, ion attraction, and electric migration. However, after calcium carbonate precipitation the solution becomes more alkaline and bucite and other crystals are precipitated out. After this, precipitation and further growth of calcium carbonate crystals are inhibited. Magnesium hydroxide bucite is the most common precipitate with calcium carbonate being usually 5-15% of the total. At resting intervals, biological organisms transform the unstructured materials into a stronger substance. The following are the precipitates which appear on submerged metal mesh: bucite, Mg(OH)2, average of 80%, aragonite, CaCo3, average of 10%, calcite, CaCo3, average of 7%, and other precipitates 3%. The cathode shape to be encrusted is a surface made up of linear wires." 1

This surface or template, for which the metal rods are reinforcing, can be fashioned into the shape of dams, dock facilities, underwater habitats, storage facilities, or ships' hulls. As accretion occurs, hydrogen, oxygen and chlorine are given off. These are then available in the seawater to buoy the structure's volume to any desired position.

Factors controlling the saturation state of calcium carbonate are variable: "Major factors controlling the saturation state of calcium carbonate in the oceans are pressure, temperature, and the CO_2—carbonic acid system. Pressure is directly proportional to the water depth. Temperature is stable in deep water (below about 1.5 km), but has significant variations with latitude, currents and upwelling in near-surface waters. In addition to temperature variations, other factors including exchange rates across the air-water interface, biological activity, and surface to deep water mixing rates can affect the carbonate concentration in near—surface waters."1

In summary, it should be noted that the technology works in most ocean areas because the most common precipitate, magnesium hydroxide, is plentiful in all oceans. The calcium carbonate precipitate obtained is mostly dependent on the precipitation technique used and is affected by the amount available in the seawater. In certain near shore regions chemicals are very concentrated and readily available.

SELF-REPAIR USING THE PEIZOELECTRIC EFFECT

Pier columns constructed of this material adapt to changes in load pressure because the structure is now under electricity-generating pressure and attracts more charged ions from seawater. It is thus self-repairing.

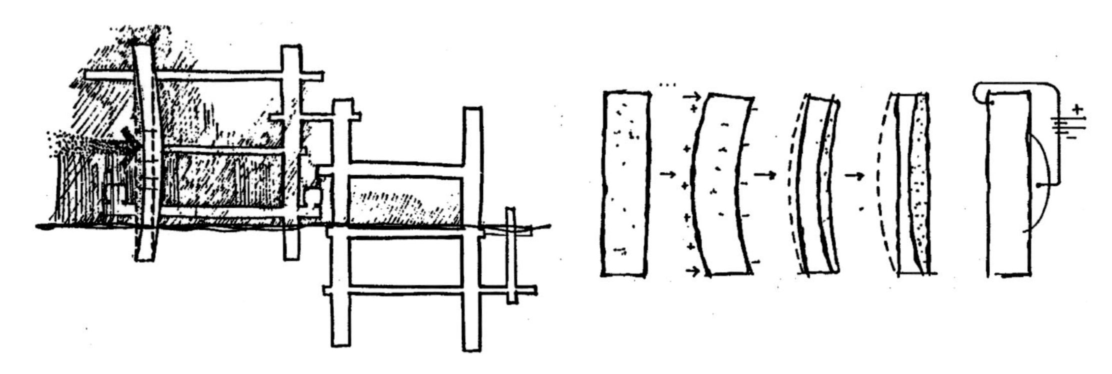

Left, diagram of pressure on column, right, diagram of the mechanism of repair of columns using piezoelectricity.5

An automatic reclamation and deposition of accreted materials to enhance the structural properties uses the piezoelectric effect of transforming mechanical energy into electrical energy and vice versa. An electrical potential develops when the material is strained due to the chemistry of the piezoelectric material. 5 It could be used to sense deformation and stress in the accreted cables or pipelines and accomplish self-reorganizing repair through the migration of electrically charged minerals to locations needing repair. This would be due to stress which generate electricity and thereby attracts ions. For instance, in bone a mechanical deformation generates an electrical current. (When a negative charge builds up on the concave side, new bone from the positive convex side fills in to straighten it.) "An electrical current applied to an undeformed bone, caused growth in the area of negative charge, no loss in the area of positive charge. The reason is that slight stress generates a charge that attracts or repels electrically charges ions in the blood plasma bathing the bone cells. Removal of stress causes reversal of charge and an opposite effect on charged ions. This electrical pumping system is responsible for translating mechanical energy into electrical energy." 5

BENEFITS OF THE SYSTEM

This is an especially exciting example of complex, self-organizing relationships among elements that is found in natural processes. This building system uses material and energy indigenous to the area and has its own inseparable installation, repair, and reclamation process integrated into one organization. The capacity to reorganize material in a complex way depends upon relationships among elements which are the results of millions of years of evolution.

The sensing function mimics nerve initiating fibers which strip off electrical charges as signals when ions flow through a nerve's charged membrane. Ions moving across a charged osmotic membrane can create an electrical impulse. This ionic impulse is the natural process way of generating current in seawater. An electrical cable in seawater is like a nerve in blood, with blood having nearly the same composition as seawater. In the human body, ionic leakage across a membrane from the potassium rich "seawater" of the nerve to normal "seawater" of the blood, can propagate a nervous impulse. The original research on nerves by A.L. Hodgkin was done by studying cables in seawater for the equations which apply to nerves in blood. 7

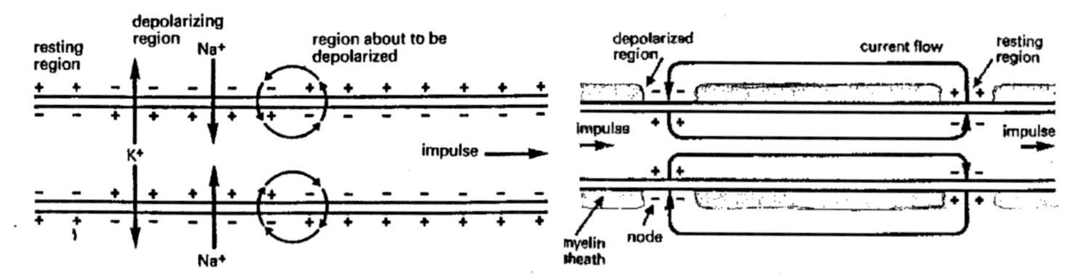

Diagrams of nerve functioning. 7

"The general conclusion is that electrical signals can be generated by the membrane without the direct intervention of metabolism, by selectively changing the ionic permeabilities and allowing ions to flow "downhill" along their electrochemical gradients. With each action, the cell gains sodium and loses potassium, but for a single impulse the amounts are small enough for the resulting concentration changes to be ignored. The poorly insulated nerve uses the leakage problems in the system to build up a large charge by the change in ionic and electrical potential across a membrane." 7 It is an example of using what materials are available and an apparent disadvantage (leaks) to tremendous advantage. Various differences affecting the chemicals in seawater such as salinity, temperature make motion possible, imitating muscles.

Reverse osmosis and electrodialysis processes are both based on the use of semipermeable membranes to achieve solute-solvent separation in saline waters. In the case of reverse osmosis, fresh water diffuses through the membrane leaving the salt behind. "In electrodialysis, demineralization of saline solution takes place by the passage of salt through the membrane. The driving force employed to cause solute-solvent separation in reverse osmosis is hydraulic pressure . In electrodialysis, electric current acts as the driving force."9

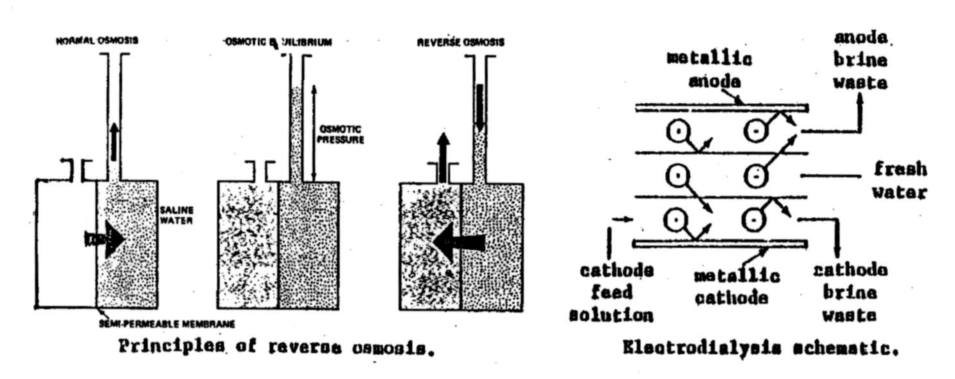

Schematics of the principles of reverse osmosis and electrodialysis. 9,12

Just how advantageous those ionic leaks across membranes can be is immense. Power (i.e. voltage and current generated directly by diffusion of ions across a permselective membrane that separates the concentrated from the dilute solution) is one of the greatest potential sources of energy in the world and equals Ocean Thermal Energy Conversion or OTEC: a process that can produce electricity using the temperature difference between deep, cold ocean water and warm surface waters.

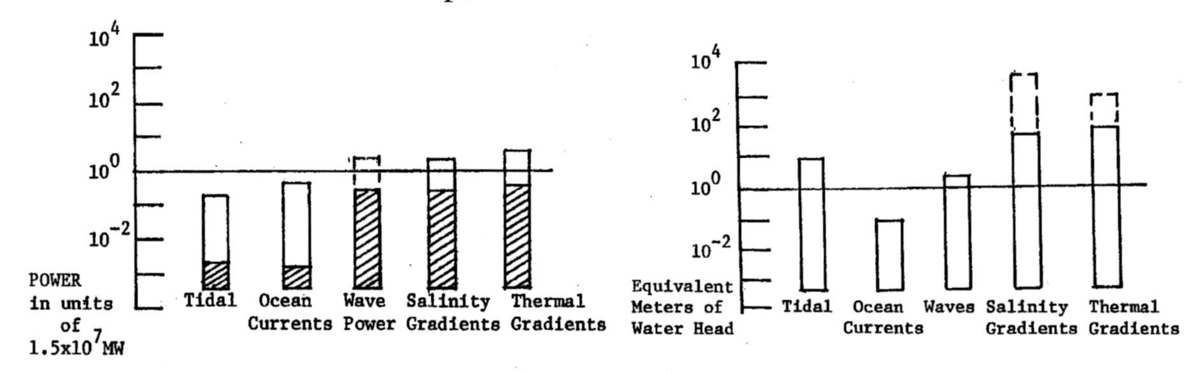

Left, power generated by various ocean processes. Right, equivalent meters of water head for various ocean processes. 10, 11

OSMOTIC PUMP CREATES FRESH WATER

A pipe open at the top but capped at the bottom by a semipermeable membrane (permeable to water but not to dissolved salts) is lowered into the ocean. "Because the osmotic pressure difference between fresh and salt water is about 23 atmospheres under ordinary conditions, the inside of the pipe will at first be free of water. When the pipe is lowered more than 231 meters the pressure difference across the membrane will exceed 23 atmospheres, and fresh water will flow into the pipe and rise to a level 231 meters below the ocean surface. Since salt water is about 3% denser than fresh water, if the pipe is lowered further the level of fresh water in the pipe will have to rise if the pressure difference at the membrane is to remain at 23 atm. Finally, if the pipe is lowered deep enough, water should rise above the surface. "[12] Thus, in principle, if you grant an ideal membrane, we should be able to get a fountain of fresh water from the ocean. This osmotic pump requires a larger membrane area and a cheap and lasting membrane system. (This pump, like the saline one, moves water against gravity, acquiring potential energy as hydraulic pressure which of course can be translated into useful energy when it is depressurized through a hydroturbine generator.)

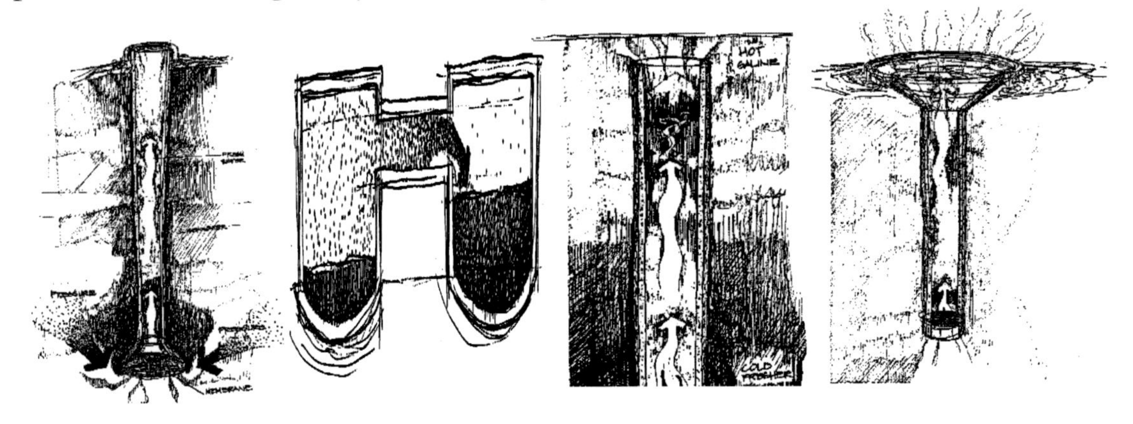

Osmotic, saline, thermal pumps and vapor pressure compensation illustrated

Building a port with natural process design becomes a symphony of water movements. The evaporative pump used as vertical port caissons, which has more saline water in it, can accrete and thicken its mineral walls most rapidly using mineral salts. The vertically moving water of different temperatures and salinity can do a number of things. It can flood or sink to evacuate the water from an area for work to be done, or float and move away on the water's surface. Once the port facility is built it continues to utilize these natural processes and water movement for various functions including dry dock facilities.

When the port is not in use, the absence of in-use pressure allows the charged ions in the structure to lose their electrical attraction and dissolve back into the seawater to be recycled. The deposited material can be reclaimed by adding chemicals and or reversing the electrode position. This is the technique used by archaeologists and marine scientists to fight encrustation. The metal wire on the surface serves as the anode. Improving the environment consists of taking up CO2 from seawater as the accretion forms and giving off hydrogen and oxygen in the process.

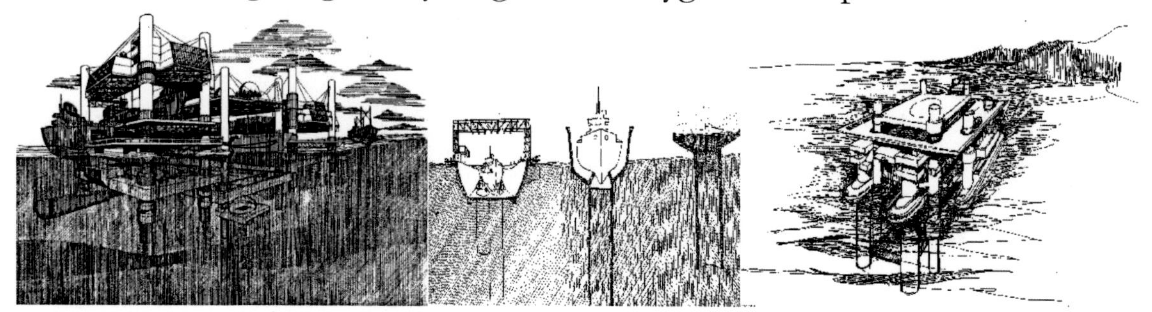

A port in which the parts can be in motion.

CONCLUSION

Available chemicals and processes in the ocean are used to accomplish port functions:

- Indigenous energy from the sun as solar energy
- Cheap materials of chemicals in seawater
- Ocean chemicals can be put into physical organization by the movement of them as charged ions in an electrolyte (capable of carrying a charge), and the seawater, by an electrical charge delivered by metal anodes and cathodes
- Self-repair uses the pressure electric (piezoelectric) quality of the structure to give off an electrical charge when force is applied such as damage. This electrical charge attracts more charged ions from the seawater to fill in damaged areas and self-repair.
- Self-sensing exterior environmental changes and communicating that information is done by charged ions flowing across a charged membrane and exchanging charges which sends a signal. This diffusion of ions across

permselective membranes separates the concentrated and diluted solutions so a voltage and current is produced.

- Salinity, temperature and osmotic differences move seawater.
- Reversible, recyclable materials are the very structural material. If you turn off the electricity or charge-producing capacity the structural chemicals eventually dissolve back into the seawater from which they came.
- Improving the environment consists of taking up CO_2 from seawater as the accretion forms and giving off hydrogen and oxygen.

Footnotes cited

1. Wolf Hilbertz, 1979 Electrodeposition of Minerals in Sea Water: IEEE Journal on Oceanic Engineering, Vol. OE-4, No.

2. 3,

pp. 94–113,1973

3. 1979, PDF 11.3 MB Experiments and Applications, in: IEEE Journal on Oceanic Engineering, Vol. OE-4, No. 3, pp.

4. 94

113, 1979, PDF 11.3 MB 1. J. P. Riley and R. Chester, Introduction to Marine Chemistry (London: Academic Press, 1971), pp. 4, 81.

5. J. W. Morse, J. deKanel, and H. L. Craig, "Saturation State of Calcium Carbonate in Seawater and its Possible Significance for Scale Formation on OTEC Heat Exchangers"–Abstract (Miami, Florida: University of Miami) p. 334,338.

4. S. A. Wainwright, W. D. Biggs, J. D. Currey, J. M. Gosline, Mechanical Design in_Organisms (London, England: Edward Arnold Press, 1976) p. 187.

5. C. Andrew L. Bassett, "Electrical Effects in Bone."

6. Lucien Girardin, Bionics. World University Library (England: McGraw—Hill Book Publishers, 1968), p. 66.

7. N. B. Gilula, M. L. Epstein, "Cell to Cell Communication, Gap Junctions and Calcium" Symposium of the Society of Experimental Biology, Number 30, Calcium in Biological Systems (Cambridge, England: Cambridge University Press, 1975).

9. K. C. Charinabasappa, "Reverse Osmosis Desalinization Technology," in Desalinization 17 (Amsterdam: Elsevier Scientific Publishing Co., 1975), p. 34.

10. G. Wick and J. D. Isaacs, "Salinity Power," Report (Institute of Marine Resources, University of California: La Jolla, California, September, 1975).

11. S. Loeb, H. R. Block, J. D. Isaacs, "Salinity Power, Potential and Processes, Especially Membrane Processes."

12. 0. Levenspiel and N. deNevers, "The Osmotic Pump," Science, 18 January 1974, pp.157-160.

CHAPTER 3:

FUNCTIONS OF THE PARADIGM: ASH BUILDING PRODUCTS AND USE OF WASTE MATERIALS

After doing the overall systemic project of ocean port using the parading and the biomimetic model of the animal body, I decided to separate out the various functions of the paradigm because I went to teach at the University of Illinois with an emphasis on engineering and specific bounded projects. I had a lab and wanted to start to prove that the ideas would work. Also by this time I had decided to focus on projects to ameliorate the excessive release of pollutants from the burning of fossil fuels. These would absorb greenhouse gases or absorb ashes from burning.

Initially at the university and then in my own small business, I did one lab-proven project per function of the paradigm.

- For recycling I used fly ash and made full size panels in cooperation with a company.
- For self-repair, I made self-repairing concrete and polymers both of which save CO_2 from being given off. Self-repairing concrete was made into full-scale bridges.
- Sensing of self-repair was done in conjunction with the self-repair projects.
- For recycling I developed pre-stressed concrete I beams that can be demolished with much less energy release from the energy stored in the pre-stressed steel tendons. In normal demolition of pre-stressed girders, the tendons can explode and destroy other parts of the structure.
- To improve the environment, I developed a coating that absorbs greenhouse gases which can also strengthen concrete if used as a coating for concrete.

In the following pages, I hope to have you able to recognize and understand the usefulness and elegance of ideas and solutions based on designing with natural processes and learn how to think about how to use this paradigm yourself with an eye to understanding where important places to intercede might be.

Currently, coal burning by utility companies produces large amounts of waste during the production of energy. Besides the carbon dioxide produced, a significant waste byproduct of coal ash waste is an environmental hazard due to its heavy metal content and carbon properties. Both bottom ash and fly ash are collected from smoke stacks because the material must be kept from escaping the plant as required by law. Collection occurs with screens above the incinerator and from below the burners. Yet even with government-regulated collection, fly ash is still a major pollutant that regularly finds its way into the environment – especially into ground water.

The use of fly ash in the building industry can be a productive outlet for the disposal and containment of this hazardous waste product. Other researchers have used fly ash as a component of several different types of construction materials including concrete, autoclaved cellular concrete flowable fill and lightweight aggregate. Fly ash also contains a portion of floaters, or air-filled spheres, which makes this material lightweight and useable for insulation. The high iron content allows for coloration changes. Melting of the spheres containing iron allows it to change color from dark rust to tan to orange to violet.

The novel aspect is to use 100% ash with only water and a flux to supply a responsible option to handle the hazardous waste of coal burning by using fly ash as the primary ingredient of various large-scale building materials. While fly ash contains heavy metals and carbon, it also has small spheres of glass which also give it unique properties that if approached creatively can be an asset instead of a liability. Also, the ashes have pozzolanic properties as a natural, low-strength cement. This innovative approach to waste makes safe and usable products while sequestering the toxic metals and ash.

Using fly ash as a building product is an example of using waste materials and the uptake of effluents of a fossil fuel economy. Some of the products made are building panels, bricks, and insulation panels. The goal is to make structural panels using the waste ash with as little other material as possible to create a cheap product and to sequester heavy metals so they do not go into the groundwater or the air we breathe.

Left, Photo of sample masonry unit made of foamed fly ash composition. Right, photo of sample structural units made into walls with metal supports.

BUILDING PRODUCTS

In the project on ash recycling to make building products, I sequestered the heavy metals in the ash by cooking or sintering 99% fly ash, which no one else has done. Fly ash is a pozzalan, a chemical that has adhesive properties like cementing chemicals. By cooking the pozzalan ash into a solid, we have another use for the ash that does not leach heavy metals into the groundwater. I used a smart choice of a flux that allows one to cook the ash at a lower temp than usual. This flux and the lower energy required uses less heat and chemical. In common practice, the ash is put as an additive into some other binder such as cement. As an additive, one can only use a small amount of theash.

I made sandwich panels of both the ashes (fly ash inside for insulation); concrete-like blocks, exterior applied insulation from the hollow spheres, a substitute for lightweight concrete using the hollow spheres, and finally large panels of the fly ash for building panels. It was found that the ash could be used with a flux to make nearly 100% pure ash building products, but that the variability of the carbon in the ash made guaranteeing the performance properties problematic. The amount of carbon left in the ash was the largest determinant of the strength, though it was too expensive to burn it all out to give uniform calcined or burned ash. The toxic heavy metals were shown to be sequestered in the products so that they would not leach out into ground water during their lifetimes of use.

Unlike other uses employed, I use nearly 100% fly ash mixed and treated with a small amount of chemical flux to reduce temperature needed for sintering into solid form. With this mixture, we formed panels and blocks using various molds and firing. Based upon our lab results, the optimal panel composition consists mainly of fly ash with boric acid, straw, ADVA Flow (a superplasticizer), and water.

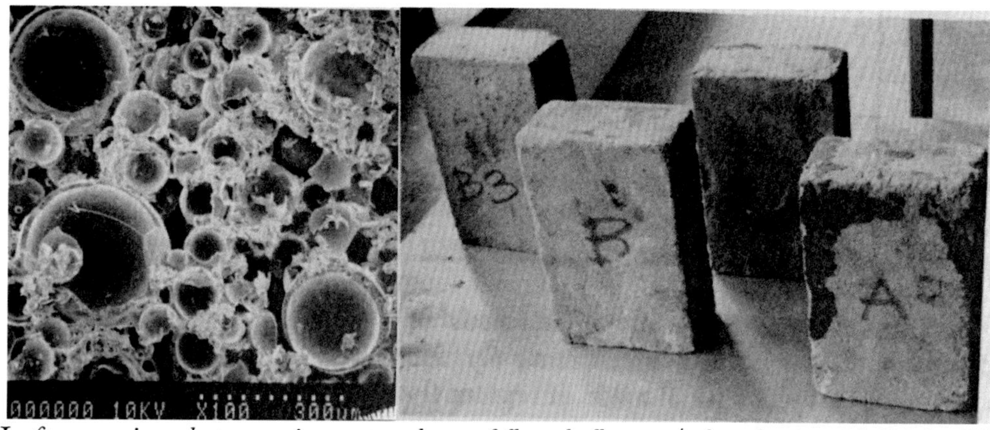

Left, scanning electron microscope photo of fly ash floaters/phosphoric acid. Right, Photo of exterior applied insulation samples made from floater fly ash material.

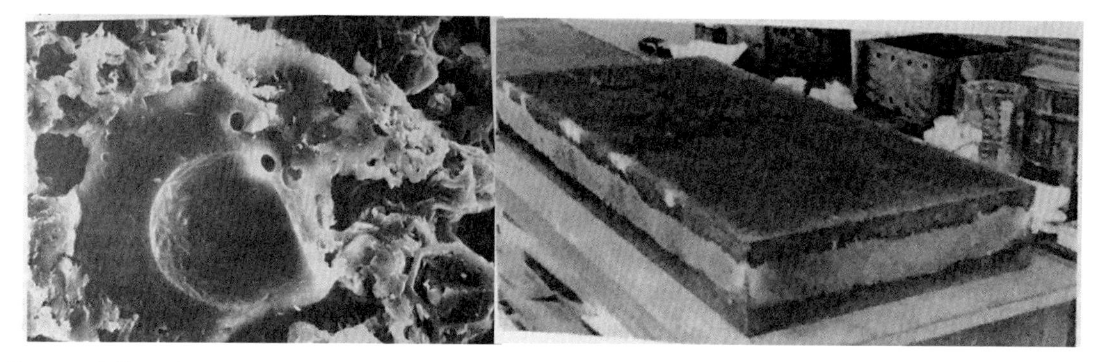

Left, Sintered fly ash containing floaters and phosphoric acid. The floaters are hollow with air so they work well for insulation, Right, a sandwich panel made of fly ash/phosphoric acid on the inside and bottom ash/phosphoric acid as the outside layer, sintered at one time.

COMPOSITION

An ash panel consists of approximately 99% fly ash. In this case, boric acid acts as the flux, increasing compressive strength and reducing shrinkage and cracking. We used straw as an inexpensive and easily attainable natural fiber to further reduce shrinkage and cracking due to drying. ADVA Flow, a superplasticizer that acts as a sort of lubricant, somewhat reduced water requirements. Increases in the temperature and duration of sintering results in impressive strength increases and decreases water permeability. We found an optimal temperature range of 800°C to 900°C for the ash compositions.

When using fly ash in any product, responsible manufacturers must address the concern of whether heavy metals could leach from their product. We performed standard leaching and nuclear resonance tests on all sample types, looking for some 30 possible chemical agents. Our results showed heavy metals well sequestered, with very little leaching particularly when sintered within the optimal temperature range (Dry et al. 2002). This proves to be an effective containment and environmentally responsible way to dispose of fly ash.

PROPERTIES OF FLY ASH

My Illinois lab used fly ash to produce samples that could be studied and refined for the material's compressive strength, thermal value, water absorption, leaching, and loss-on-ignition characteristics. It is well known that fly ash is a pozzalan and has cementitious properties of its own. We made small scale samples of bricks, cubes, sandwich panels, and blocks composed entirely of fly ash not only for simplicity, but also to optimize sintering behavior, reduce shrinkage and cracking due to drying, and to enhance insulative properties and mechanical properties, particularly compressive strength. Ash from coal-fired furnaces was made into sandwich panels and building systems, foamed and used as hollow spheres to be an alternative to cellular concrete, exterior applied insulation, and full-scale panels.

This research was done under the Advanced Construction Technology Center sponsored by the U.S. Army Research Office and its goal was to produce a material which would be structural as well as low-cost, lightweight, and insulative for building components made as sandwich panels or blocks. The performance parameters were structural capacity, high strength-to-weight ratio, inexpensiveness, casting of chases into the material, low temperature sintering, good insulative value, flexure strength, and durability.

After evaluating the many waste materials available and processing variations, we chose ash materials which meet these parameters and use just a few common chemicals (acid and ashes) which bond the materials together for a sandwich panel. We also chose low-temperature sintering for chase in-casting. Compositions were optimized, then the two materials were combined. A mixture of fly ash, alumina, and acid was developed and evaluated as a lightweight foam interior material with a low firing temperature. Bottom ash and acid were developed and evaluated as a strong exterior material with a low firing temperature. Structural sandwich panels were then made from bonded and sintered (cooked) bottom ash on the outside with bonded and sintered fly ash on the inside. The fly ash balloon floaters or hollow spheres give good insulative properties and make the material lightweight. The strong (3,000 psi) bottom ash exteriors lend ease to sandwich panel construction. Metal chases for electrical wiring or pipes cart be cast in because the materials are sintered at a low temperature of 600° C, so the metal will not melt. Close packing of various particle sizes gives improved impermeability and as well as the insulative properties.

Sintered foamed fly ash and acid composition and fly ash floater and acid material as alternatives to foamed or cellular concrete was the material made of an energy waste product that has been developed for uses similar to cellular concrete. It uses hollow ash spheres or can be foamed using alumina. The foamed sintered (600° C) fly ash/alumina/acid composition is lightweight, 800 kg/m3, thermal conductivity 0.0062 W/mK, and of moderate compressive strength 5000-7000 kPa. A fly ash material made of spheres was then used to improve insulative and weight properties. This material's improved performance has a lower density and a higher insulative value and higher compressive strength than concrete. It has a microstructure of large discontinuous spheres or hollow pores, while concrete has many smaller pores uniformly distributed.

The foamed fly ash/acid alumina material is quite similar to cellular concrete in that the origin of the air voids is a reaction aluminum which produces hydrogen gas. The typical properties of insulating cellular concrete are density 300-1100 Kg/m3; compressive strength 300-7000 kPa, thermal conductivity one to 0.3 W/mK [2]

Two structural materials made of fly ash and acid showed that compressive strength can be moderately increased and lighter density and the good thermal properties are maintained as in the floater fly ash material. Development of a densified honeycomb cellular fly ash material using hollow spheres was researched. Theory shows that this is the way to increase compressive strength, yet maintain a low weight and insulative properties.

The potential use of fly ash materials in exterior insulation and finish systems (EIFS) used the same two types of fly ash insulation material: foamed fly ash and hollow fly ash spheres. These materials have a higher compressive strength than the insulation presently used in EIFS systems and so could have better durability. The objective of the research was to assess the properties of insulation material made from fly ash waste. Insulation material was developed from bonded/sintered fly ash waste material and assessed as a substitute for the polystyrene insulation board (EPS and XPS) used in conventional EIFS systems. It was hypothesized that fly ash material would provide a greater resistance to mechanical damage while realizing a thermal value which approximated that of the insulation materials traditionally used in EIFS claddings.

Exterior insulation and finish systems (EIFS) are non-load bearing, insulative cladding systems that have been in use in this country for more than eighteen years. The common types of insulation used in EIFS claddings are expanded polystyrene (EPS) board and extruded polystyrene (XPS) board. The insulation component is attached with adhesives or mechanically fastened to the substrate. A base coat material, typically reinforced with an embedded fabric for strength and rigidity is applied over the insulation board. EIFS clad- dings are typically finished with an acrylic top coat.

The study assessed the two types of fly ash compositions. The first consisted of a mixture of fly ash waste material combined with water, acid and alumina. The mixture was foamed and then sintered at 600° C. Initially, a calcia-acid bond is formed, but is replaced by corundum as the temperature goes above 350° C. The resultant foamed fly ash is a lightweight material, 800 kg/m3, with a thermal conductivity of 0.17 W/mK at a mean temperature of 24° C. Foamed fly ash has a moderate compressive strength of less than 000 kPa. A second type of fly ash material was produced with hollow fly ash spheres. The hollow fly ash spheres were mixed with acid and sintered at 600° C to produce a compar- atively lightweight material, 520 kg/m3, with a thermal conductivity of 0.11 W/mK at a mean temperature of 24° C. The compressive strength of the hollow fly ash material is 5000-7000 kPa. In some experiments, EPS beads were added to the hollow fly ash material to further reduce its weight. Table 1 provides relevant data on the fly ash materials and conventional types of insulation used in EIFS assemblies.

Microscopic examination of these materials helps explain the differences in compressive strength, density, and thermal values between the two fly ash sintered materials (foamed arid hollow).

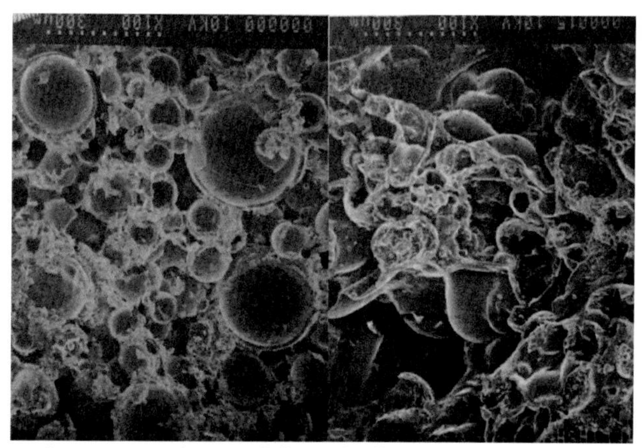

An electron microscope photo of fly ash floaters with phosphoric acid

This research was funded through the Office of Coal Development and Marketing and the Illinois Clean Coal Institute (ICCI).

In another project sponsored by the Illinois Clean Coal Institute the objective was to develop commercially viable building products at a ceramics company using Illinois coal ash from a coal-fired power plant. The technical approaches to achieving this objective were: to demonstrate the feasibility of sequestering ash metals in a value-added product with appropriate structural, insulative, weight, water absorption, and color properties; to refine the microstructure in order to optimize the performance parameters of strength, density, water resistance, and insulative properties, and to refine the production parameters at an industrial facility to produce full-scale building panels. The energy cost of sintering the panels was not considered as a factor because the heat needed could be generated on-site at the coal-fired furnace plant.

The research presented here investigates the utilization of an Illinois class F fly ash as a primary constituent of large-scale building panels. Like I said, ash contains heavy metals and is radioactive, which makes it a most difficult ash of which to dispose. Therefore, an outlet for this waste product is needed.

I have invented a novel way of using only fly ash and a flux plus water to make building wall panels and investigated the viability of this ash material for use in building products including loss on ignition studies of ash correlated with strength data, thermographic analysis to measure interior heat soak rates, specimen composition experiments, and studies of sintering regimes. The top priority concern for use of this ash is leaching of heavy metals into the environment by water action.

For the laboratory studies portion of this project, Illinois fly ash samples were produced to study and refine the compressive strength, thermal value, water absorption, leaching, and loss on ignition characteristics of the material. Small scale samples of bricks, cubes, and blocks composed entirely of fly ash mixed with some water were made to optimize sintering behavior, reduce shrinkage cracking, and enhance insulative properties, and mechanical properties, particularly compressive strength.

Compressive strength results show the samples that recorded the highest compressive strength were those that were cast with calcined ash and 10% boric acid with a three-day sintering regime at 700°C, with a resulting strength of 19.1 MPa (2770 psi). In regard to sintering temperature, increasing the temperature to 800°C was found to result in strength gain, where strength nearly doubled for cast samples with 10% boric acid and was over 7 times higher for samples cast with 10% glass as the flux.

These investigations found that calcining ash increases the compressive strength, provides the lowest water absorption, and allows for the lowest water to solids ratios necessary for casting. The presence of boric acid in the composition as the flux results in stronger samples than glass powder compositions. Increasing the sintering duration and temperature increases compressive strength and decreases the water absorption (although 800°C appears to be the upper limit before strengths begin to decrease with increasing temperature). Tests also showed calcining the raw ash prior to making the sample results not only in uniform interior cross-sectional sintering, but also results in a sample with considerably higher strength and little water absorption,

Fly ash batch	**Water / Solids**	**Boric acid**	**Ave. Max Stress (MPa)**	**Ave. Max Stress (psi)**	**Loss on Ignition at 550°C (loss of carbon)**	**Loss on Ignition at 900°C (end loss of carbon)**
FA1	.23	10%	21.0	3043	+2.75%	+7.5%
	.28	10%	14.0	2029		
FA2	.23	10%	1.57	228	+1.5%	+23.0%
	.28	10%	1.33	193		
FA3	.23	10%	13.8	2000	+4.5%	+8.4%
	.28	10%	16.3	2362		

Results of compressive strength tests and LossOnIgnition on three batches of fly ash.

Sample	**R-Value** (Ft^2Fh/BTU)	**K-Value** (W/mk)
10% glass powder, 0.25 water/solids, cast	7.626	0.0386
10% boric acid, 2% carbowax, 0.10 water/solids, compacted	7.209	0.0408

Results of insulation tests on laboratory fly ash bricks.

Block	**Water / Solids**	**Composition**	**Mechanical Test**	**Max. Stress (MPa)**	**Max. Stress (psi)**	**% Water Absorption**
1	.32	Cast Non-calcined ash 10% boric acid ADVA Flow straw	Compression	1.79	259	33.3%
			Flexure	0.224	32.5	
5	.22	Cast Calcined ash 10% boric acid ADVA Flow Straw	Compression	5.98	866	11.6%
			Flexure	0.311	45.1	

Composition and mechanical testing of laboratory fly ash blocks.

Panel	**R-Value** (Ft^2Fh/BTU)	**K-Value** (W/mk)
10% Boric Acid, cast	5.934	0.0496
10% Glass Powder, cast	7.449	0.0395

Results of insulation tests performed on brick samples from panels.

The top priority concern for use of fly ash is heavy metal leaching by the action of water. Standard leaching and nuclear resonance tests analyzing for some 30 chemicals were performed on all the sample types. The results showed that some heavy metals were well sequestered, particularly when sintered. The best sequestered and least-leached chemicals are chromium and lead. However, arsenic is leached as are some other metals. This proves to be an improved way for disposing of fly ash as compared to dumping it is mines where all metals leach out in water.

	1) Percent Composition of Raw	2) 0% acid, 3hr	3) 30% acid, 3hr	4) 0% acid, 3hr	5) 30% acid, 3 hr fire	6) 0% acid, no fire	7) 30% acid, no fire	Method Detect
Element		Concentration of						(m2/L)
As	0.0512	3.15	0.794	0.904	1.53	0.27	0.058	0.038
B	0.0930	14.8	700	17.5	1353	50.5	1166	0.014
Ba	0.0406	0.167	0.018	0.068	0.040	0.054	0.040	0.002
Be	0.00181	<0.002	0.004	<0.002	<0.002	0.312	0.060	0.002
Cd	0.0100	0.290	0.429	0.029	0.593	8.46	1.43	0.003
Co	0.00587	0.006	O.D75	<0.003	0.102	0.715	0.111	0.003
Cr	0.0314	<0.004	<0.004	<0.004	<0.004	0.611	0.100	0.004
Cu	0.0272	0.109	0.131	<0.003	0.028	1.14	0.293	0.003
Fe	10.9	0.181	3.54	0.101	1.07	4.94	2.37	0.009
M2	0.473	76.7	56.2	32.0	125	156	25.5	O.D78
Mn	0.0317	1.48	3.13	0.138	5.86	9.65	1.55	0.002
M	0.0364	7.23	0.223	14.6	0.707	<0.006	0.012	0.006
Ni	0.0209	<0.007	0.106	<0.007	0.226	6.02	0.705	0.007
Pb	0.218	<0.028	<0.028	<0.028	0.116	0.128	0.068	0.028
Se	0.038	<0.034	<0.034	<0.034	<0.034	0.077	<0.034	0.034
Si	3.69	6.62	12.2	7.9	13.5	15.0	5.99	0.034
V	0.0521	0.994	0.025	2.96	0.239	0.023	0.013	0.002
Zn	10.9	0.300	6.09	0.031	5.40	371	52.6	0.002

Results of composition of raw ash (col. 1) and leachate analyses from fired ash mixtures (cols. 2-5) and unfired ash mixtures (cols. 6-7).

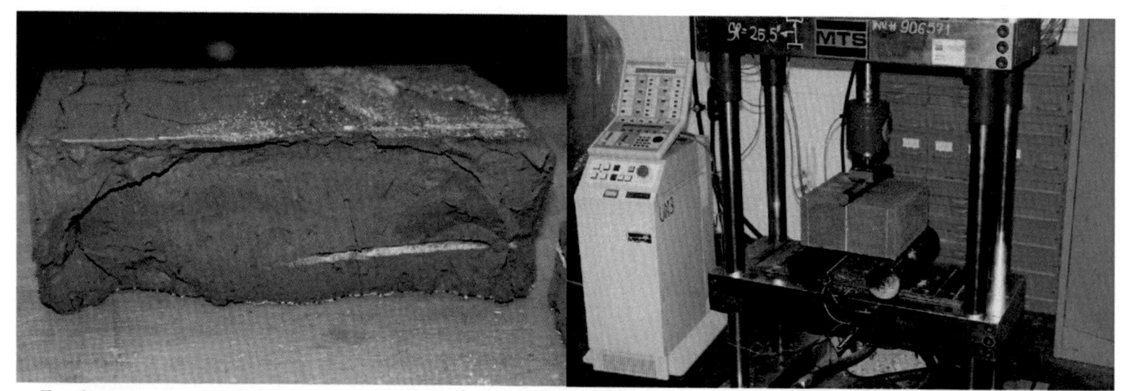

Left, interior of laboratory test block (calcined ash, boric acid, superplasticizer, straw) demonstrating uniform sintering. Right, testing a block in compression.

From the results of the small-scale tests, full scale panels were produced and tested.

Based upon results and observations of the small scale pour and work done in the laboratory, boric acid was included in all panels of the second pour and the further use of straw and metal fibers to reduce cracking of the specimens was considered. In addition, superplasticizers were investigated that could reduce the water required for adequate casting. Although laboratory investigations demonstrated that calcining the ash prior to mixing and making into products reduced the water requirement while providing the highest strength, it was determined that calcining on a large scale was not economically feasible due to energy costs. It also results in problems which include clumping of the ash due to calcining and shrinkage cracking. Prior lab work and panel pours proved that straw could be used to reduce drying shrinkage and cracking.

The full-scale panels were fired at 815° C (1500° F) for ten hours and exhibited better sintering depth than previous ones. This was most likely due to the longer firing time. A considerable amount of shrinkage occurred, but panels remained very dense and heavy.

Left, cast full-scale panel in steel mold prior to sintering. Next, full-scale panels after sintering at 700 degrees Celsius. Right, full-scale panel fabrication at industrial facility.

Left two, casting of full-scale panel, Right, soil compactor used in some panel's fabrication.

Panels made with glass powder exhibited a higher insulating R-value than those made using boric acid only. In addition to the thermal resistance value, the panels also allow for a long thermal lag which can allow for higher energy efficiency in building applications.

For future research, the following suggestions should be considered.

- Calcining the ash would solve problems with the lack of sintering depth by providing a uniform, smooth, and hard cross-section while increasing compressive strength and reducing water absorption and water requirements. Since the amount of carbon present in the same type and source of ash can be variable among batches of ash received from
- the power plant, calcining would also maintain consistency among the batches by burning out remaining carbon.
- After firing and testing a sample made with a hammer press at the factory, it was seen that compacted samples could yield strength and stability if compacted enough. If more sufficient methods of compaction on a large scale could be employed, then compaction may prove to be viable.
- Improvements of insulative properties could be attained through vacuum sintering of the ash panels.
- The creative use of waste materials reduces environmental impact of our fossil fuel use and makes a marketable product out of waste.

CHAPTER 4:

SELF-REPAIRING CONCRETE

The manufacture of cement generates an incredible 8-10% of the world's CO_2. The goal of this inquiry was to extend the service life of concrete structures so less cement needs to be produced.

Concrete is a very common material but is brittle, porous, and reacts to intrusion of water and chemicals from the environment All of the conventional solutions for preserving concrete address making it less porous or preventing damage once the environment intrudes (anti-corrosion and anti-cracking, etc.) by adding chemicals and treatments later.

Our natural process design solutions have materials built in before the concrete is formed to reduce the porosity and prevent the various types of damage that occur over time as well as repair damage brought on by cracking of the brittle material. This is important because of its widespread use and the huge impact it has on the environment.

Specifically, our research took place in three parts. One exploits the biomimetic aspects by using the bone analogy of flowing repair chemicals from inside the body of the concrete itself. A second part remedies each of the problems brought on by porosity in lab scale experiments. The third attempts to scale up solutions for corrosion and cracking damage into full-scale bridge tests.

- Streaming and piezoelectric repair based on inherent properties of the cement itself to flow repair chemicals available from inside the cement to damage sites based on an under- standing of bone repair actions.
- Inclusions to reduce intrusion of damaging chemicals and water into porous concrete and designs specifically to reduce each of the associated damages such as corrosion, freeze thaw and the related porosity.
- Inclusions of self-repairing chemicals in hollow vessels such as fibers embedded in the concrete to repair cracking and corrosion damage with testing done on large scale bridges

SELF-REPAIR BY STREAMING POTENTIAL AND PIEZOELECTRIC EFFECT IN CEMENT

Bone is a material after which to model construction materials for many reasons. Bone is strong, tough, and adaptable including its great strength, toughness, and adaptability. Bone has the intrinsic ability to adapt to its environment, namely loading conditions. Research on the body's electrical properties reveals that two phenomena occur to allow bone to adapt to environmental changes: the inherent piezoelectric nature and streaming potential. Together they create charge differences that attract ions to specific regions of the bone, namely those under greatest stress, in order to build up the region to handle the applied load. We researched the utilization of these properties in cement to increase adaptability. We studied the inherent electric properties of the cement itself and considered the introduction of a different polymer or ceramic within the cement to impart piezoelectricity and streaming potential so that it could self-repair.

Bone exhibits the piezoelectric effect: an electric potential that develops when the bone is strained. Bending in bone can be opposed by building up bone to curve in the opposite direction, therefore the applied bending moment will serve to straighten the bone. Area of bone that are in tension (convex) are broken down, and areas that are in compression (concave) are built up.

Piezoelectric signals from these strained regions call for appropriate action in the bone. Bone uses this piezoelectric affect to attract or repel local calcium ions in order to build up or break down the bone in a given vicinity. This builds up the bone to support applied loads optimally.

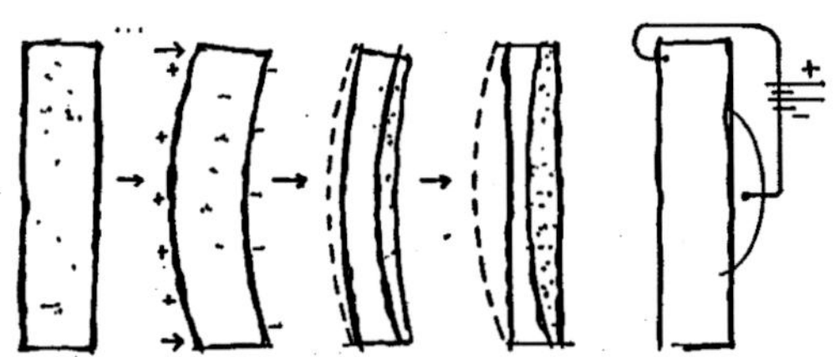

Frost's model of adaptive bone reconstruction: A) a piece of unloaded bone is subjected to B) axial load and bending moment, creating (c) concave areas of compression and (v) convex areas of tension. C) The bone is remodeled by piezoelectricity, building up concave areas and breaking down convex areas (dotted areas = old bone) When loaded again it will straighten. 1

In this research, consideration was made of utilizing these piezoelectric properties in cement. They could help control the formation process or allow cement to adapt and strengthen due to environmental and loading changes. The cement itself might have (or be made to have) inherent piezoelectric properties of enough significance to exhibit these piezoelectric properties, which could be of benefit during cement formation and lifetime adaptation.

In the bone analogy, self-repair by use of the streaming potential would use an electric potential difference that exists within bone, which is caused by the streaming of an electrolytic fluid through some porous media such as bone. More specifically, the streaming allows bone to self-repair. "Liquid flows though the stressed porous bone and negative particles in the liquid are attracted to the positively charged ions on the bone's pore surfaces. As the liquid loses its electrons to positively charged ions on the bone surface, it continues to flow through the bone with an increasing abundance of the positive charge. The net result of this very rapid stripping of some charged ions and deposition in another location is a difference in charge (a potential difference) between the two areas of the bone. This allows positively charged ions to be drawn from the tension or positive region of the bone ("downstream") and be deposited "upstream" to build up the compression or negative region of the bone. The streaming potential is different from the piezoelectric effect but closely related to it; it is thought to be perhaps partially responsible for the effect of piezoelectricity. " 2

This streaming potential which is useful in bone might be utilized in this cement as well. By forcing an electrolyte solution to pass through porous cement, the streaming potential would cause a charge difference in the stressed composite thus causing the charged particles to be deposited to strengthen the stressed area.

As a practical application, bending members would squeeze the electrolyte solution from the compressed area ("upstream") to the tensile area ("downstream"). Therefore, charged ions would buildup in the compressed region of the bent section. This would help by shifting the centerline of the member toward the compressed side.

Streaming potential is caused by liquid being forced through a stressed cement column (by bending of column and the pressure of enclosing) and depositing charged ions on the compression side which yields the piezoelectric remodeling of column as noted in Frost's model.

Dynamic loading conditions offer the best opportunity for use of these adaptability mechanisms of streaming potential and piezoelectricity to reshape the cement composite material according to the stress fields. To this end, dynamic testing will be done. The cement was dynamically loaded and any potential difference was measured.

Samples from the leg bone of a cow were immersed in a saline solution. Cement samples were prepared as small round samples with average thickness of approximately .5 cm. Wendell Williams et al. investigated the intensity of streaming potentials in the mineral and cellular portions of bone. To conduct these experiments, they developed an instrument in which a porous material could be held in place, and a sodium chloride solution was forced through it, while the electrical potential across the sample was measured. In our experiment, a nearly identical experimental setup was used except for the thickness of the samples. The sample was fastened in place in a tube-shaped polycarbonate apparatus, which was then filled with salt water.

Electrodes extended into the solution and were connected to a multimeter. The plungers were driven by an Instron testing machine at 200mm/second. A potential with an absolute value approaching ten millivolts was produced each time the plungers were moved and the salt water was driven through either the bone or cement sample. This differs from the results obtained by Wendell Williams et al. where a potential of roughly one tenth that magnitude was measured. When no sample was present in the apparatus, a change in potential was not produced as the plungers were moved.

Streaming Potentials

Sample Type	Initial Potential	Potential During Flow	Absolute Potential
Bone	12.0µV	142µV	30µV
Bone	114.5µV	116µV	1.5µV
Cement	123µV	123.5µV	0.5µV
Cement	94.7Mv	92.5Mv	2.2µV

CONCLUSIONS

The ability of bone to adapt to its environment, namely loading conditions, piezoelectric properties, and streaming potential were considered as models, because they are the basis for this type of adaptability. Promising results found that a streaming potential is produced in bone or cement when a sodium chloride solution flows through it, but the potential is slightly higher in bone than cement.

In this project (done for my PhD dissertation at Virginia Tech University) I recognized that concrete, a "strong and permanent" material, changes drastically over time and the environment seriously degrades it with corrosion, drying and cracking, alkali silica reactions, freezing and thawing damage, and other chemical processes. These forces of deterioration could instead be the signal for regeneration by following the natural process design paradigm.

Costs for maintenance, repair, and replacement for structures and infrastructure is much higher than initial costs. As a result, durability has become one of the important cost issues for design. Conventional designs for components and structural systems address the durability mainly by the depth of the concrete; to prevent corrosion a thicker cover is needed and to reduce micro-cracking fibers are included. However, durability is not a function of the concrete material by itself. It is the relationship of aggressive environmental agents intruding and ingressing into the material. This relates to the material properties, especially permeability, which controls the rate of ingression and the rate of penetration of destructive environmental agents which deteriorate material properties. The deterioration of concrete is closely related to the way in which aggressive environmental agents penetrate and attack. Because permeability controls this rate and the associated pore structure, size, distribution, and cracking, it is a significant measure of the ability of environmental agent's intrusion.

Durability must be evaluated as a function of performance versus time. It is as influenced by the environment in which the concrete structure serves as well as the specific properties of the concrete itself. Durability problems specific to concrete in its environment must be addressed by changing its adaptable properties over time.

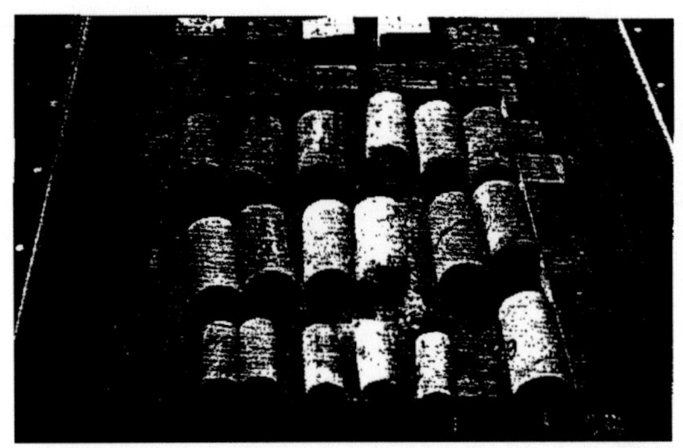 

The test set ups for permeability and freeze thaw damage

The hypothesis of this study is that by incorporating physical or chemical means into the matrix of the material for later release, the material matrix can evolve or adapt to changing environmental and material circumstances over time; parameters such as matrix permeability, pore make up, and crack structure can be altered. This can alter the potential for environmental degradation such as freeze/thaw and corrosion damage.

DESIGN OF SYSTEMS FOR ACTIVATED INTERNAL RELEASE

The release of chemicals into the matrix to adapt the material to changing circumstances over time can be in direct response to the intrusion of the environment. The intrusion of chloride ions, the intrusion of water, which can freeze, or loading which can cause cracking are examples of this. A planned response to changes in the life of the concrete can also enhance durability.

Examples of release in direct response to the naturally timed intrusion of the environment are: crack sealants may be released from porous fibers flexed due to loading, calcium nitrite, an anticorrosion chemical, may be released from porous polypropylene fibers near the rebars due to degradation of a pH sensitive fiber coating of polyol which degrades at pH 11.5 (the level at which corrosion starts, thus reducing the corrosion on the rebars), porous aggregates saturated with propylene glycol may release these chemicals into the matrix when freezing/thawing pressures occur, thus the freeze/thaw damage may be reduced, and Xypex, a catalytic crystallizing agent, may release from prills, after the matrix is set. A prill is a mixture of Xypex and wax or polyol made into capsule form. When heat is applied the wax melts (or it releases the Xypex due to polyol degradation in alkalinity), releasing the Xypex to crystallize in the matrix. This reduces matrix permeability.

These materials are designed to address the general durability issues of permeability (porosity, structure, size), cracking due to loading and self- desiccation and the specific penetration, attack of chloride ion intrusion (causing corrosion), and water and ion intrusion subjected to osmotic pressure causing damage such as freeze/thaw damage.

If a chemical is released internally, then it could move more easily and quickly throughout the concrete matrix giving better coverage, or bonding with the cement. If a chemical is released over time in response to environmental events, then it would reduce the time-dependent degradation factors such as freeze/thaw damage, corrosion damage, and so forth, by responding to each event as it occurs. If a specific environmental event triggers the remedial action required to prevent damage from that event within the concrete, then the performance parameter could be increased, and the cost of the material designed for protection from that environmental event reduced.

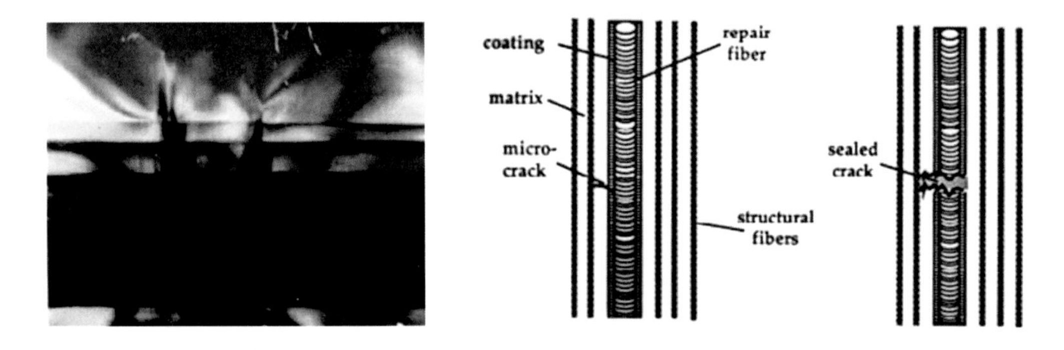

Left, schematic representation of the self-repair mechanism. Right, A photo, taken under a microscope, of a coated glass pipette releasing its repair chemical into the matrix, under tensile stress.

The following research questions were answered for specific designs for timed internal release of chemicals: Cracking due to loading is a major problem in concrete materials because the cracks increase permeability. Loading over time has a cumulative effect which can lead finally to complete deterioration of the component or structure. The design to alleviate this problem consists of filling porous fibers with a crack-filling chemical. The chemical is released from the fibers when the fibers flex due to loading. The stimulus for release is thus loading. This is the ideal situation in which the agent of environmental degradation is the stimulus to release the repair chemical. The testing regime design relies heavily on the release of chemicals from the fibers due to bending caused by loading, therefore the research or investigation focuses on that area. Another problem was the ability of water and chemicals to intrude because the matrix is permeable is the major problem in durability.

Permeability is addressed by the release of sealants into the body of the matrix from hollow, porous fibers. The methyl methacrylate, the repair chemical, is contained in the fibers for later release by coating the fibers with wax. To be effective, the methyl methacrylate must be released from the fibers after the matrix is set up and then polymerize. Heat is the stimulus which not only releases the methyl methacrylate from the porous fibers, but also dries out the matrix to receive the methyl methacrylate and polymerizes it in place. The heat stimulus has that threefold effect: accomplishing release, matrix drying, and polymerization. The investigation in this design focused primarily on the increase in impermeability but also on increased compressive strength and increased bending strength added by fibers.

Permeability is the main durability problem because that is the way in which environmental agents penetrate to attack. In this design, Xypex, a catalytic crystallizing agent which crystallizes in the presence of water and causes the 25% unhydrated cement to hydrate, is released from prills. These prills are paraffin wax in which the Xypex is mixed with stimulus for release being heat. In another design using Xypex, the prills are made from a polyol which is alkaline sensitive and should degrade due to the stimulation of the alkalinity change in cement to more alkaline as a precursor to corrosion. The research investigated increased impermeability as well as any attendant increase in compressive strength.

The problem of corrosion can be approached by reducing the ability of the steel to corrode. Calcium nitrite added freely to cement protects the steel but also retards the concrete set time. In this design calcium nitrite is put into hollow, porous fibers coated with wax and placed near the rebar. The rebar is heated, releasing the chemical from the nearby fibers. Also, the fibers could be coated with polyol. The stimulus for that release is the change in alkalinity to 11 at which point corrosion starts. This stimulus has the advantage that the mechanism of attack, i.e., the reduction of alkalinity to 11 due to the ingress of chloride ions. This is also the stimulus for activation of the prevention system. The research in this investigation focused on reduction in corrosion as measured by reduction in differences in electrical potential across rebar.

The problem of freeze/thaw damage is important because it breaks up the concrete internally, causing cracking and increased permeability. The design in this case is to load porous aggregates with chemicals which reduce freeze/thaw damage. These chemicals should be released due to the expansion of water and the changes in osmotic pressure in the specimens upon freezing. In this case, the stimulus is the agent of environmental attack (freezing). The chemicals to be released are linseed oil, a hydrophobic agent that reduces the ability of water to penetrate internally (an anti-freeze), and propylene glycol. The anti-freeze should reduce freeze/thaw damage by decreasing the temperature at which freezing will occur.

Does the successful release alleviate the environmental distress or reduce cracking or permeability? These answers are different for the different systems.

Cracking caused by repeated loading was tested with a bending apparatus (a load cell). Preliminary test results pulling the filled fibers under a microscope showed that the polypropylene fibers stretched instead of forcing out the chemical. Porous fiberglass fibers broke before forcing the chemical out.

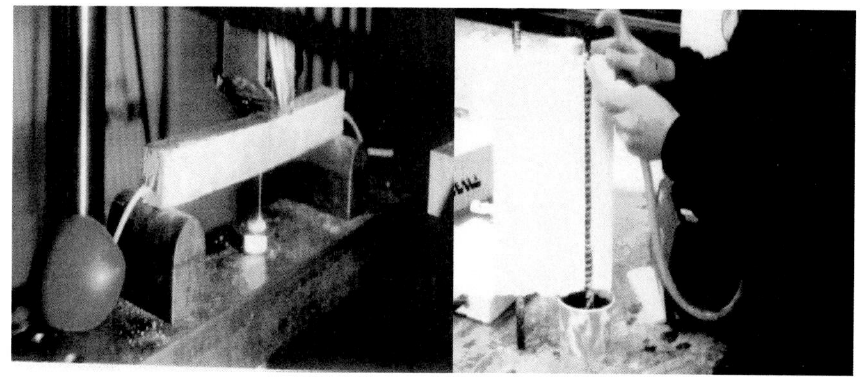

Left, demonstration of the pressure release concept. Adhesive in the fiber and a balloon is under pressure. When the specimen cracks, the pressure differential causes the adhesive to flow rapidly into the crack site. Right, demonstration of the vacuum concept, filling of embedded repair fibers.

The design to resist chemical or water intrusion due to impermeability uses methyl methacrylate contained in polypropylene fibers coated with wax. It was tested by the gravity flow permeability apparatus (Ludijdsa, Berger, and Young, 1989). The results show that methyl methacrylate in polyprophylene fibers, but not polymerized (heated only to 120°F), decreases the permeability. Methyl methacrylate released and polymerized (heated to 212°F) reduces the permeability more and has lower permeability than the control.

The test for increased impermeability using Xypex prills was done with the gravity flow permeability test also. Results in show that the addition of Xypex/wax prills does not increase impermeability.

The design to resist corrosion of reinforcing bars in concrete by release of calcium nitrite from polypropylene fibers coated with wax was tested by the time of setting of hydrated cement by Vicat or Gillmore needles ASTM C191-82. This design does not retard the set as much as the freely mixed calcium nitrite.

Freeze/thaw damage was assessed for 6"x 3" cylinders using the standard ASTM C 666 test using a Logan freeze thaw machine which cycles 300 times. The conclusion is that time release of antifreeze and linseed oil somewhat reduces freeze/thaw change in some cases.

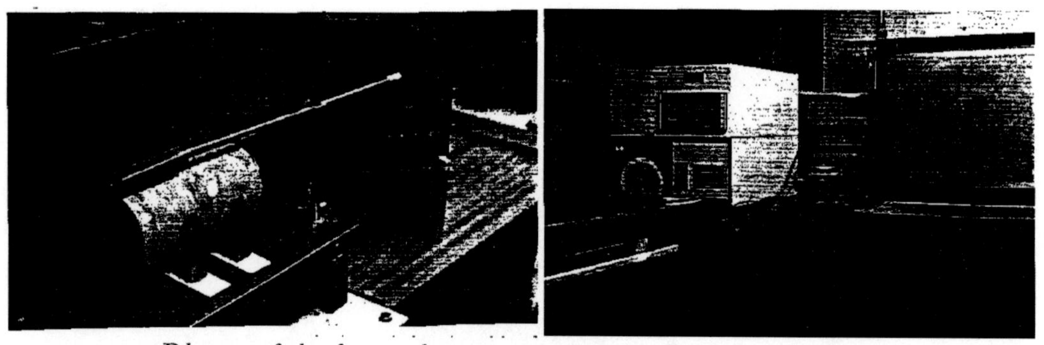

Photos of the freeze thaw test equipment Logan machine.

Are these designs an improvement over conventional treatments for impermeability or crack reduction and environmental distress? The internal release of methyl methacrylate from hollow, porous fibers was compared with the conventional polymer impregnation surface application of methyl methacrylate and inclusion of fibers. The same amount of methyl methacrylate was added on the top of the sample as was contained in the fibers. Bending and compression strength tests and permeability were done. Surface application was done at the same time the internal release samples were heated. Methyl methacrylate contained in fibers were stronger in both bending and compression than the samples in which methyl methacrylate was applied from on top (as is done in the field). Permeability was also better for internal release.

One of most successful examples of preparing the matrix to receive the repair chemical and doing the repair was heat melting the wax coating to allow methyl methacrylate (or calcium nitrite) to release from the fibers and heat drying the matrix in preparation.to receive the methyl methacrylate (or calcium nitrite).

In the design to address corrosion, the internal release of calcium nitrite was compared with the inclusion of calcium nitrite in the initial mix. The comparison for set retardation ASTM test 191-82 was used. The results show little set retardation with calcium nitrite filled fibers but significant set retardation with calcium nitrite freely mixed in the cement.

In the case of linseed oil or antifreeze released internally to address freeze/thaw damage, the comparison was made with some air entrained samples. The results show less weight loss damage for the internal release and less length loss damage if internal release is combined with conventional. Success is due to freeze and thaw action driving the propylene glycol or linseed oil from the porous aggregate into the matrix.

In the case of a catalytic sealant—Xypex to reduce permeability—the comparison was made between the sealant released from wax prills contained inside the samples with the sealant applied externally, both with and without wax particles in the cement. The surface application was done at the same time the internal release samples were heated. The comparisons were made for permeability, bending and compression strength. The bending and compress ion strength were greater for Xypex wax prills in the samples than the same amount of Xypex applied on top of the sample, as is done conventionally in the field. Permeability was approximately the same for internal release.

INITIAL CONCLUSIONS

Tests were made to evaluate the effectiveness of the successful combinations for the treatment of particular distresses. It was concluded that methyl methacrylate released from fibers and then polymerized in the matrix increases impermeability and that Xypex did not increase permeability. Also, freezing damage was reduced somewhat by the release of chemicals during freezing. Calcium nitrite in fibers did not retard the set which is the desired result (actual corrosion tests were too lengthy for this research study). Conclusions drawn from tests in comparison with conventional treatments were that the internal fiber release of methyl methacrylate is an improvement over the conventional external polymer impregnation. Xypex results were better for the timed release than conventional applications. Freeze/thaw damage, as measured by weight loss, was less with the system for internal timed release of chemicals than with air entrained concrete, but not as good if measured by length change as those combining the internal release and conventional treatment. Calcium nitrite in fibers does not retard the set at all, while conventional, fully-mixed chemical does retard the set. Storage of filled fibers seems to present no problem if closed containers are used to prevent chemical evaporation. Longer time in field testing of all factors is needed, as well as a study of manufacturing issues.

The anticipated major cost and design implications of internal timed release are that a large part of life-cycle costs can be predicted as part of the design cost. The time to first repair would be extended, and therefore the annual costs reduced. In these designs, the repair time begins with environmental penetration; the time when normal deterioration begins. In the cases of human stimulation, the repairs begin when the stimulus is given, which is before any environmental penetration and deterioration occurs.

Usual approaches for repair of structural concrete are polymer injection, pre-stressing, geomembranes, and polymer wraps. These techniques seek a ductile, less brittle failure. All of them are based on addition of a repair material to concrete from the outside in; we add the materials from "the inside out" of the concrete for self-repair.

Our approach consists of embedding repair material in hollow fibers in the repair matrix, before it is subjected to damage. Therefore, when cracking occurs, this repair material is released from inside the fibers and enters the matrix, where it penetrates cracks and re-bonds to the mother material of the structure being repaired. The cracking and damage, associated with the low tensile strain capacity, triggers the release of the repair material. This is important because in this way the material acts in an extensible manner. Further, it does crack and show ductile behavior, but the instant repair of cracks and damage assures that deterioration does not accompany this "crack then repair" behavior. The problem is repaired where it occurs, just in time, and automatically without manual intervention.

"Long-term durability is achieved by dimensional stability, which means less stress from thermal contraction, autogenous shrinkage, and drying shrinkage. The combination of factors affecting crack resistance is called "tensile strain capacity" or "extensibility." Cracking of concrete can be managed by controlling the extensibility of the material. Cement-based materials with large extensibility can be subjected to large deformations without cracking." The technique we utilize does precisely that. The process adds more material to the concrete matrix from the inside upon demand when it is triggered by events such as cracking of the matrix or shrinkage. Moreover, it does not sacrifice strength for durability since the fibers and released repair adhesive add to the overall strength.

Various techniques have been proposed for repair, however all of them have been only partial and temporary solutions because concrete as a material is brittle. Also, concrete structures are dimensionally unstable because of movement. Consequently, the usual repairs do not hold. The technique we have developed, self-repair, unites the most crucial qualities necessary for a successful repair system: crack resistance and long-term durability. This approach addresses the bonding problem of repair material from inside the concrete, therefore, it is a better technique compared to other methods such as polymer injection. Furthermore, compared to the use of compressible beads in polymer injection, it is seen that self-repair performs better because the adhesive is flexible itself and keeps on releasing with each brittle failure, namely a crack. The ability to fill in for dimensional gaps has been shown to work with self-repairing adhesives that foam. Even with internally released stiff, non-foaming adhesives such as cyanoacrylates, self-repaired matrices are less brittle, more ductile, and stronger in tension than controls without adhesives.

The question of whether we could replace the tensile strength given by steel rebar was explored. The research results showed that in the first loading, concrete samples with adhesives and a small amount of metal that provides additional tensile strength were stronger in tension than control samples reinforced with metal wires, metal mesh or even rebar. In self-repair of cracks, different types of structural failure require different approaches.

For instance, in full scale bridge applications, we addressed surface drying cracks by creating in-situ control joints. We placed scored brittle fibers, which broke at the centerline upon matrix shrinkage, releasing a sealant/adhesive. Furthermore, we repaired shear cracks by chemicals released from fibers buried in the depth of the bridges. *We observed that when these broke, the entire bridge performed better than the control one without the embedded adhesive filled fibers.* Our approach provided a self-repairing technique that transformed the entire structure into a ductile material where energy was dissipated all over as cracks formed, and consequently, catastrophic failure due to the enlargement of any one crack was prevented.

Furthermore, we made frames to represent bridges and structures subjected to dynamic loads. We demonstrated that because of self-repair with different types and locations of adhesives there was less permanent deflection, more stiffness in specific locations where desired (by adding stiff adhesives), and more damping when adding damping chemicals.

Our approach consists of embedding repair material in hollow fibers in the repair cement matrix before it is subjected to damage. Therefore, when cracking occurs, this repair material is released from inside the fibers and enters the matrix, where it penetrates cracks and re-bonds to the mother material of the structure being repaired.

Further, we addressed the corrosion problem by the release of anticorrosion chemicals from hollow porous walled polypropylene (fiberglass) fibers which were coated with a chemical that dissolves in saltwater, namely polyol. The encapsulated chemical is released onto the metal rebar when the coating is dissolved by the saltwater.

THE PROBLEM: DAMAGE IN CONCRETE, REPAIR EFFICACY AND COST

Concrete is inexpensive relative to other construction materials. However, damage can greatly reduce its life cycle with internal damage being common in concrete. Repairing this type of damage is crucial in preventing failures that can progress to catastrophic ones. However, it is hard to detect micro-scale cracks unless they have developed to macroscopic scale flaws. Non-destructive evaluation techniques have limited ability to detect these micro-cracks. Also, the damage repaired in the field by hand does not restore the original strength of the material.

Concrete materials have applications in rehabilitation of existing bridges. They are either used for complete structural replacement or new construction. The American Association of State Highway and Transportation Officials (AASHTO) projected that to maintain current bridge conditions, 200,000 bridges will need to be replaced or repaired during the next two decades. The main problem is quality assurance.

200,000 bridges are projected to need large-scale maintenance in the next 20 years. CC Image courtesy of Graham Soult via freeimages.com.

Deterioration of concrete structures is another challenge that we are faced with. The problems in concrete for civil structures are: n repair and survival of damage, n reliability comprising over time repairability, durability and health assurance, and n cost, but especially life cycle costs of repair or replacement. We have focused on improvement of reliability due to damage, durability, and design of low-cost concrete repair systems for structural use which have low life cycle costs, and lower initial costs. Various repair chemicals were used which could n repair different types of damage n fill different sizes of damage n and repair damage caused by different forces and speed.

This research produced a new family of self-healing concrete repair materials, which will solve the problems mentioned above and relieve the fears of planners. Self-healing solves the quality assurance problem and reduces life cycle costs. Extended life reduces

the number of replacements, and future costs. This new family of self-repairing materials is less
expensive overall because the repair material is built in and available wherever and whenever it is needed. Therefore, over-designing for damage protection is eliminated. The life cycle cost (usually much more than initial construction costs) will yield the most dramatic savings.

SELF-REPAIR SOLUTION: THE ANSWER TO DAMAGE IN CONCRETE

"In concrete and other cement-based materials, microcracks already exist at the interfaces of the aggregate-mortar and reinforcement-mortar. When large, visible cracks become interconnected with microcracks, the network of cracks facilitates the transport of aggressive ions and gasses to the embedded reinforcement, leading to premature corrosion and deterioration." This research addressed the repair of micro damage by the release of chemicals from fibers into matrix microcracks, so that development of further damage can be prevented. The repair fibers also release chemicals between delaminating layers or re-bonded fibers to the matrix. Two different repair mechanisms for self-repair have been investigated: release from brittle fibers and release from brittle coating over porous walled fibers. Experiments assessed the ability to both re-bond fibers and repair cracks using fiber pull-out tests, impact, and bending tests with successful results.

We investigated the repair of microcracking damage, fiber debonding, matrix delamination, fiber breakage, impact damage such as holes, and adhesive debonds and their healing by the release of chemical healing agents from fibers into the matrix. These damages are all affected by interfaces. Reduction of damage in the matrix is done by release of repair chemicals from fibers directly into the interfaces.

To be self-repairing, a healing chemical is stored in hollow fiber vessels embedded in the polymer matrix. When the composite is damaged, the crack progresses, breaking the repair fiber. The healing chemical flows into the crack and re-bonds the cracked faces. Alternatively, the fiber can be re-bonded to the matrix and to delaminations and holes repaired with the adhesive.

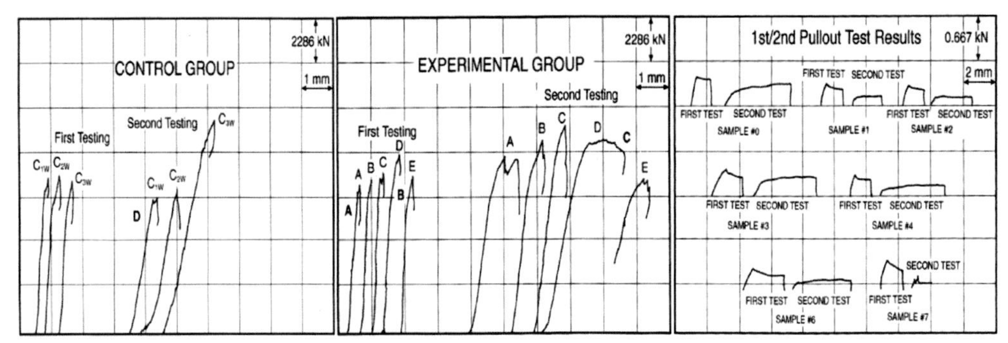

Example of strength increase and more ductile failure due to repair chemical in small concrete test beams, left. Center, the repair fibers were subjected to pull out testing and found to be re-bonded after release of the adhesive into the matrix. Right, the pullout tests show a more ductile release with the fibers that had been adhesively rebonded than the controls.

DESIGN OF THE SELF-REPAIRING SYSTEM

We made self-healing repair material consisting of embedded, continuous hollow glass fibers containing waterproof adhesives that filled cracks when and where they occur. Continuous fibers with a high aspect ratio gave the best composite properties in stiffness and strength. Further, as the volume of fibers increase, the stiffness and strength of the structural system increases. Up to 80% volume fraction of fibers is acceptable if the fibers can be incorporated into the matrix. To resist corrosion, porous, walled polypropylene fibers coated with polyol, which dissolves in salt water, were used. These contained an anti-corrosion chemical and when the polyol dissolved, the chemical was released onto the metal rebar.

In general, materials capable of passive, smart self-repair consist of several parts: an agent of internal deterioration, such as dynamic loading, which induces cracking; a stimulus to release the repairing chemical; a fiber or bead; a coating or fiber wall which can be removed or changed in response to the stimulus; a chemical carried inside the fiber; and a method of hardening the chemical in the matrix.

TIME RELEASE IN SMART MATERIALS

After developing several examples of time release in smart materials, I drew up the following list of attributes desirable for a successful design optimization of such a system for different applications.

- The types of problems or distress that can be effectively treated by chemical release must be determined. The attributes of those types of distress are: a) There must be a time dependent problem of durability relating the exterior environment to the matrix conditions. b) The problem must be important in cost and a frequent cause of deterioration. c) The distress could be treatable by the chemical released.
- The internal stimulating agent could be chosen if needed to allow the matrix to accept the chemical to be released, or provide other chemicals (for crosslinking).
- The release mechanism from the fiber should be designed for the specific application.
- The types of chemicals to be released, i.e., the encapsulant, must be effective in treating the specific type of distress.
- The physical properties of the release agent or encapsulator, that is, material and shape should be tailored to the treatment of the specific environmental distress, and to other needs of the design of the component. The method of encapsulating chemicals into the fiber or microcapsules must be an efficient and inexpensive process.
- The stimuli or source of energy or change must be such that it can cause the chemical to be released from the fibers.
- The matrix must be such that it can accept the encapsulant in form, chemistry and volume and accept the chemical as it is released and accept the stimuli action necessary to release the chemicals.

RESEARCH ON SELF-REPAIR OF CRACKS IN CONCRETE

We worked on "butter stick" samples that were 1"x 1"x 6" in size. We placed 24 small metal-reinforcing fibers and 150 ml glass pipettes which were filled with adhesive in each sample. The control samples did not contain any glass pipettes, but they were reinforced with 24 small metal reinforcing just like all the other samples. After tests, we found that in members subjected to bending, the failure mode was more brittle in the first break and more ductile in the second break, especially when compared to the control ones. Pull-out tests also revealed that the fibers were re-bonded to the matrix, after the glass pipettes were broken and the internal adhesive released.

In further tests, the ability of adhesives to fill in for dimensional gaps has been discovered. The self-repairing adhesive that foamed were the most successful ones. This result was very important because it showed that although concrete was a non-extensible material, it could behave in an extensible way should the right choice of self-repairing technique be pursued and adhesive types utilized.

Four reinforcing systems in standard mix concrete composite matrices were investigated under bending. The question of whether we could replace the tensile strength given by steel rebar was explored. The research results showed that in the first loading, concrete samples with adhesives and little metal for tensile strength were stronger in tension than control samples reinforced with rebar, metal wires, or metal mesh.

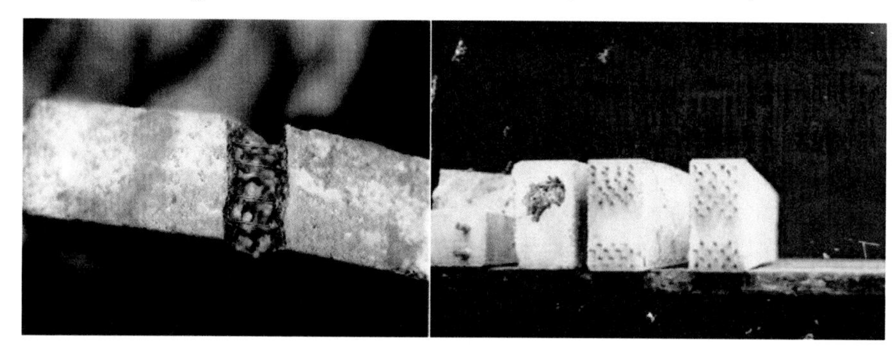

Left, a photo of a beam repaired by foaming repair adhesive. Right, testing results comparing four reinforcing systems under bending in standard mix concrete composite matrices.

Description at 2.5 mm. deflection	**Test 1 (kN)**	**Test 2 (kN)**	**% of Change in load**
Wire Only	22.0, 10.5	16.0, 0.0	-27, -100
Reinforcing Bar #4	12.5, 8.5	8.0, 0.5	-36, -100
Steel Fiber Mat	5.5, 11.0	16.0, 0.0	-97, -100
Wire & Adhesive	19.0, 9.0	19.5, 1.4	-3, -84

Finally, we compared our results of internal self-repair where we addressed the bonding problem from inside the concrete to the polymer injection method which was done from outside to inside. It was shown that the self-repair technique fared better than the polymer injection method. Furthermore, compared to the use of compressible beads in polymer injection and regular polymer injection into brittle cracks, self-repair was again better in repairing because the adhesive itself is flexible, and kept on being released with each crack.

Sample Type	**Modulus of Elasticity**	**Cracking Load Increment (%) Average Values**
Results or polymer injection from outside	High	56*, 65*, 83, 102
	Low	103
Results of internal self-repair from inside	High	121, 123
	Low	88, 119, 132

SELF-REPAIRING CONCRETE FRAMES WHICH REPRESENT BUILDINGS AND BRIDGES

The self-healing method investigated for this project utilized the timed release of adhesive into the member at the time of cracking. We constructed concrete frames that simulated buildings and bridges. Chemically inert tubing was cast within the cross section of the member and then was filled with adhesive. At the onset of cracking, the tube wall was fractured, allowing adhesive to exit the tubing and penetrate the developing crack. Two sets of static loading were applied, in which failure modes were checked for each sample to determine either the frame failed at crack sites sealed by the flexible adhesives, or crack sites sealed by adhesive. Then the frames were subjected to cyclic loading (repetitive static loading, not dynamic) immediately after the third test in order to examine whether or not each experimental adhesive was able to exhibit elastic or inelastic behavior in the frame.

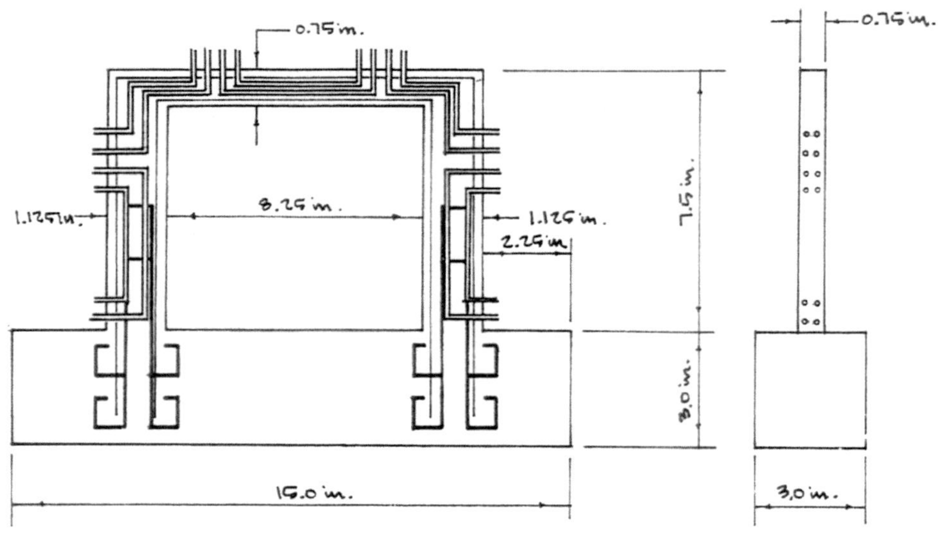

Design of test frames with glass tubes containing adhesives at joints and element midspan.

It was seen that in self-repairing frames, high absolute value of elasticity adhesives (stiff adhesives) released at the structural points repaired the initial damage in critical regions. These stiff adhesives allowed damaged points to regain stiffness, preventing future damage at the joints while transferring forces to other portions of the structure, preventing catastrophic failure. However, the control frames with no internal adhesive were catastrophically damaged.

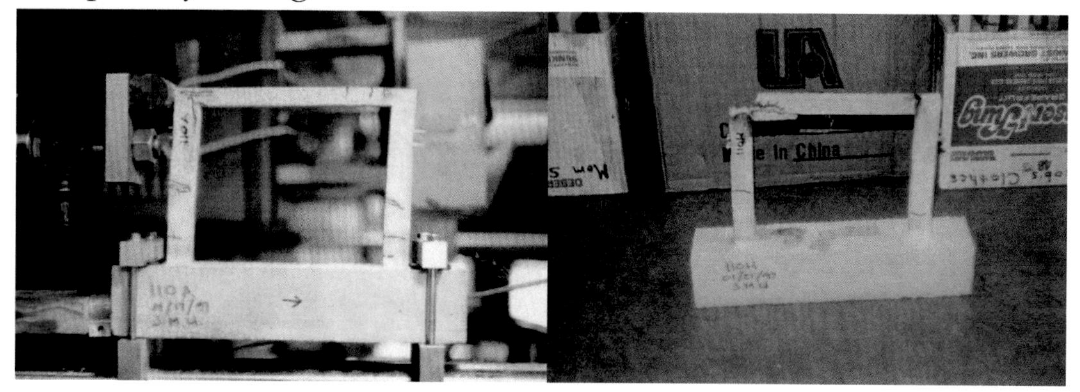

Left, a photo of frame containing internal self-repairing cyanoacrylate adhesives, cracking in the third static test, which shows cracking all over, but no catastrophic failure. Right, a photo of control frame with no internal self-repairing adhesives, which shows catastrophic failure after testing.

The self-repairing frames deflected more than the control ones while resisting larger loads. Self-repairing frames had fewer reopened old cracks than controls, and the self-repairing frames with stiff adhesives were stiffer than the controls–the ones without adhesives. Furthermore, testing on the hysteresis effects (phenomenon in which the value of a physical property lags behind changes in the effect causing it, showed that the frames with repair chemical more nearly returned to their original configuration that those without repair chemicals.

BENDING TESTS AT 3MM		BENDING TESTS AT 0.6 KN	
Sample	% Increase	Sample	% Increase
Control	-52.38	Control	20.00
101b	-45.45	101b	71.43
101c	-9.09	101c	29.03
101d	16.67		

Thus, we successfully demonstrated the control of structural damage by strategic release of appropriate internal repair adhesives in critical locations of the frame. There was less permanent deflection and more stiffness in locations where adhesives were used and more damping where damping chemicals were added to the matrix. It was proven that structural damage, namely cracking, can be directed to the members themselves, where cracks can be repaired by flexible adhesives which allow flexibility in the members for energy dissipation. This is necessary for resisting dynamic loading failure and recovery from deformation. The most interesting result of these experiments was the visual assessment in transparent frames that the adhesive was being pumped further with each crack opening and closing due to reloading.

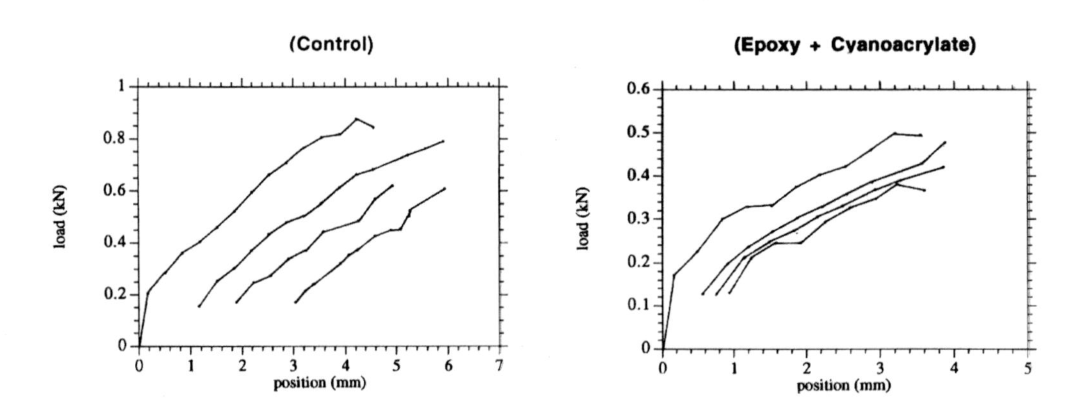

Hysteresis test results on self-repairing frames on right, left frames without repair chemical

SELF-REPAIR OF CRACKS IN FULL-SCALE BRIDGE DECKS

Different types of cracking require different approaches. This research, sponsored by NCHRP of the Transportation Research Board and Natural Process Design, Inc. and done at the University of Illinois, focused on the repairing of drying shrinkage cracks and repair of structural load-induced cracks in four full-scale bridge decks. I was the principal investigator on this project. As an initial test, fibers were thrown into the cement mixer to prove that they could survive the mixing intact.

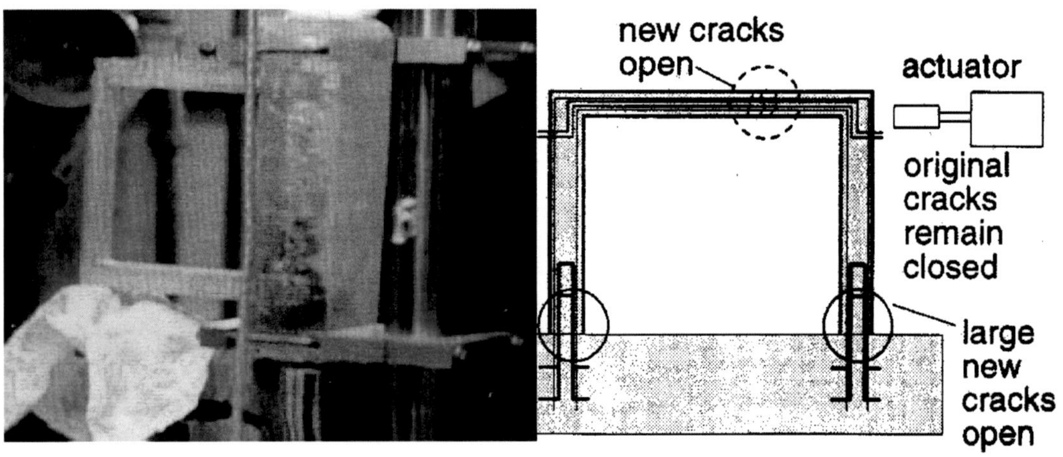

Test frames contained glass pipette fibers filled with adhesive. Not only do such frames repair cracks all over the matrix, but also it was observed that crack opening and closing (caused by load application and removal) drove the adhesive deeper into the matrix with each action, like a bellows.

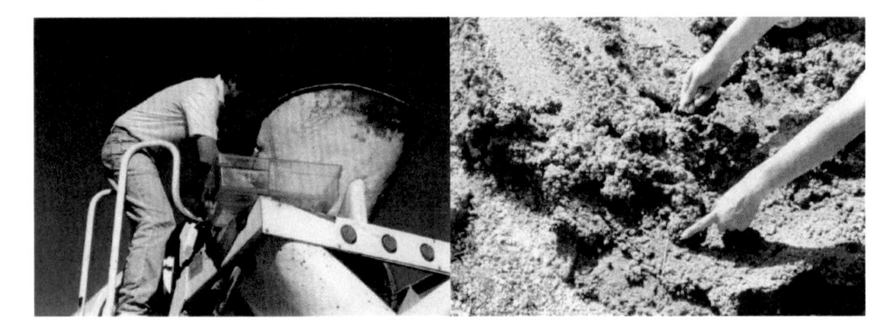

Fibers being thrown into the cement mixer, left, they survive after mixing in the bridge concrete, right.

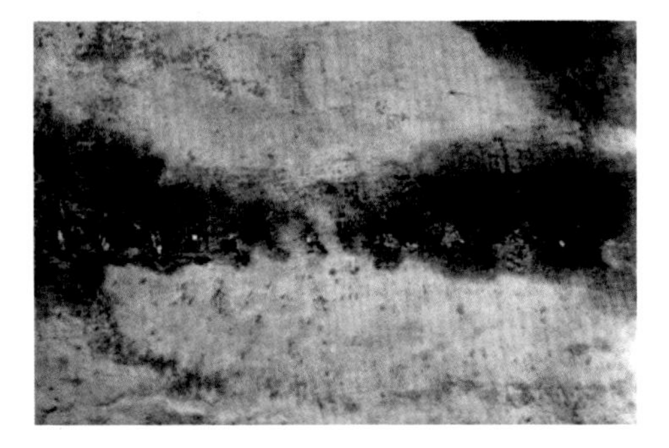

Pattern of dots of adhesive/sealant released from scored fibers make an in-situ control joint in response to drying shrinkage

We addressed surface drying cracks by creating in-situ control joints. Scored brittle fibers which broke at the centerline upon matrix shrinkage were placed, releasing a sealing adhesive. Also in these large bridges, interior shear cracks were repaired by chemicals released from fibers buried in the depth of the bridges. Long capsules containing strong, high modulus adhesives were placed below the surface in areas of tension caused by bending; for example, in the top of the section over supports. When these broke, the entire bridge performed better than the control one without embedded adhesive-filled fibers. Structural cracks induced by loading were successfully repaired. This was evidenced by the higher strength of the bridge decks that contained adhesives when compared to the control deck. We also observed that new cracks formed in certain locations and consequently prevented reopening of any of the previously repaired cracks.

The most impressive evidence of the structural crack repair capabilities of this system are the diverted cracks in the second loadings of two decks that contained adhesives. In both cases, original cracks from the first loading were repaired and secondary cracks opened—at least in portions— during the second loadings before the primary cracks did reopen. The control deck without adhesive had no such occurrence. Compared to the second and third loadings of the control deck, deck #1 (which contained no repair adhesives), decks #2, #3, and #4 all showed signs of bending strength re-gain in their later tests.

In all the decks containing repair adhesives, subsequent loadings revealed additional adhesive release along the reopened cracks. These adhesives survived for over three years in field conditions ranging from below freezing to over 100°F in central Illinois.

Regarding these results, it is seen that our approach provided a self-repairing technique transforming the entire structure into a ductile material where energy was dissipated all over as cracks formed. Consequently, catastrophic failure due to the enlargement of any crack was prevented. Cracks were repaired as they formed, so that further crack damage in those locations by the intrusion of water or chemicals was prevented at that site.

Photo of fabrication of bridge decks. Right, Photo of released repair adhesive

	Mod. of Elasticity (ksi) Test 1 Load 1-2	Mod. of Elasticity (ksi) Test 2 Load 1-2	Mod. of Elasticity (ksi) Test 3 Load 1-2		Stiffness (kip/in) Test 1 Load 1	Stiffness (kip/in) Test 2 Load 1	Stiffness (kip/in) Test 3 Load 1		Strain Test 1 Load 1	Strain Test 2 Load 1	Strain Test 3 Load 1
Deck 1	2133	853	2276	Deck 1	4.712	4.400	1.257	Deck 1	.000039	.000039	.000137
Deck 2	4267	1707		Deck 2	3.141	1.100	1.675	Deck 2	.000059	.000156	.000117
Deck 3	1067	569	3413	Deck 3	0.785	5.656	4.400	Deck 3	.000313	.000039	.000039
Deck 4	1067	5120	3413	Deck 4	0.785	3.144	3.140	Deck 4	.000313	.000039	.000078
	% Change Between Test 1 Load 1-2 and Test 2 Load 1-2	% Change Between Test 2 Load 1-2 and Test 3 Load 1-2	% Change Between Test 1 Load 1-2 and Test 3 Load 1-2		% Change Between Test 1 Load 1 and Test 2 Load 1	% Change Between Test 2 Load 1 and Test 3 Load 1	% Change Between Test 1 Load 1 and Test 3 Load 1		% Change Between Test 1 Load 1 and Test 2 Load 1	% Change Between Test 2 Load 1 and Test 3 Load 1	% Change Between Test 1 Load 1 and Test 3 Load 1
Deck 1	-60	167	7	Deck 1	-7	-71	-73	Deck 1	0	251	251
Deck 2	-60			Deck 2	-65	52	-47	Deck 2	164	-25	98
Deck 3	-47	500	220	Deck 3	621	-22	461	Deck 3	-88	0	-88
Deck 4	380	-33	220	Deck 4	301	0	300	Deck 4	-88	100	-75

Results of testing of bridges

Testing revealed that internal release of adhesives in the three tests increased the modulus of elasticity, stiffness, and reduced the strain of the three decks, as compared to the control deck.

CONCLUSIONS

Long capsules containing strong, high modulus adhesives were placed below the surface in areas of tension caused by bending. Structural cracks that were induced by loading were successfully repaired as evidenced by higher strength than a tested control deck without adhesives and by the creation of new cracks in places where the old repaired cracks had not reopened but remained repaired. This did not occur in the controls. The bridge decks were loaded 4 times over 3 years. Again, these adhesives survived for over 3 years in field conditions ranging from below freezing to over 100°F. The strength gain and/or behavioral changes as well as stability of the encapsulated repair chemicals were assessed as viable.

RESEARCH ON PREVENTION AND DELAY OF CORROSION

"Deterioration and distress of repaired concrete structures in service are a result of a combination of physical and chemical processes, such as the corrosion of embedded reinforcing steel, alkali-aggregate reaction, delayed ettrringite formation, etc. These processes are accelerated by the cracking of the repair materials, thus allowing the ingress corrosive elements such as water, salts, carbon dioxide, sulfates and oxygen, into the concrete" [8].

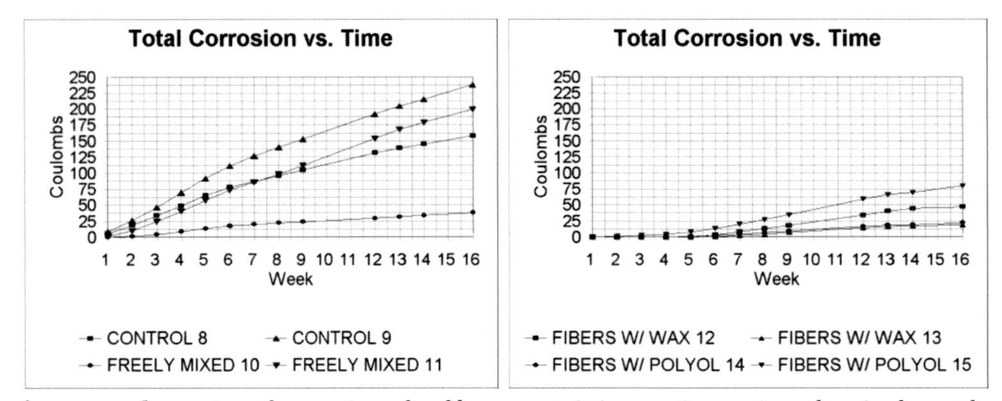

Comparison of onset and severity of corrosion, for fibers containing anticorrosion chemical, on the right, versus those without, on the left.

The time of corrosion onset and severity can be improved by the release of anticorrosion chemicals from hollow porous polypropylene or fiberglass fibers. These fibers are coated with a chemical which dissolves in saltwater, polyol. The encapsulated anti-corrosion chemical is released onto the metal rebar when the salt water dissolves the coating. We have demonstrated this method in the laboratory using ASTM tests for corrosion. The release of the corrosion inhibitor chemical is at the portion of the reinforcing bar in danger of corrosion when conditions would allow corrosion to initiate. In a series of tests with concrete samples containing either no protection or the conventional freely mixed calcium nitrite, this system of internal release from fibers performed well. It delayed the onset of corrosion by at least three weeks in the laboratory specimens and reduced the amount of total corrosion by more than half.

Samples on which a sponge soaked in various anticorrosion chemicals in placed near the rebar and then an electrical charge with a negative charge, is applied to drive the positively charged ionic chemical out of the sponge and towards the metal rebar, which is based on similar concept for bridges by Sam Hettarachi.

SELECTION OF CHEMICAL ENCAPSULANT

Chemicals should be selected which will (a) flow out of the fiber into the matrix, (b) are effective in a matrix with a particular crack size and structure, delamination, hole or fiber re-bond issue, (c) alleviate the distress, (d) do not interfere with the other properties of the matrix, (e) have a low enough viscosity to enter the repair fiber initially and flow into the cracks, and f) do not set up under heat of hydration. We have successfully used one-, two-, and three-part systems. We made samples for testing which contained methymethacarylate/cumine hydroperoxide initiator and cobalt noedconate system, a three-part system.

Photo of our samples for testing which contained methymethacarylate/cumine hydroperoxide initiator and cobalt neodeconate system, a three-part system.

References cited

1Frost model of bone repair, . Wainwright, S.A., W.D. Biggs, J.D. Currey, and J.M. Gosline, Mechanical Design in Organisms, Halsted Press, New York, 1976.

2Streaming potential . Ibid

CHAPTER 5:

SELF-REPAIRING POLYMERS

Airplanes that are safe to fly even if damaged, bridges that can withstand an earthquake and remain standing, boats that can resist damage to keep afloat, and oil rigs and pipelines that can resist damaging storms are the contributions of self-repairing composites to the economy and to social welfare. New planes and ships are being made from polymers in order to be lighter and therefore more fuel-efficient. Ships and plane are hugely responsible for global warming in that most cargo is sent by ship and by plane which both give off large amounts of carbon dioxide and chlorofluorocarbons. This causes global warming and damage the ozone layer. By using lightweight, oil-based polymer composites that delaminate and break easily, planes are prone to undetected damage that can be catastrophic. Manufacturers don't trust them so they make them thicker and heavier which reduces the lightweight advantages. Self-repair eliminates these worries and is estimated to reduce the weight by 30% in airplanes. The goal of my wing project was to extend the service life of polymers composites and make them much lighter to save vast amounts of fuel while extending their durability, thus using less oil-based product and making potentially safer polymer vehicles.

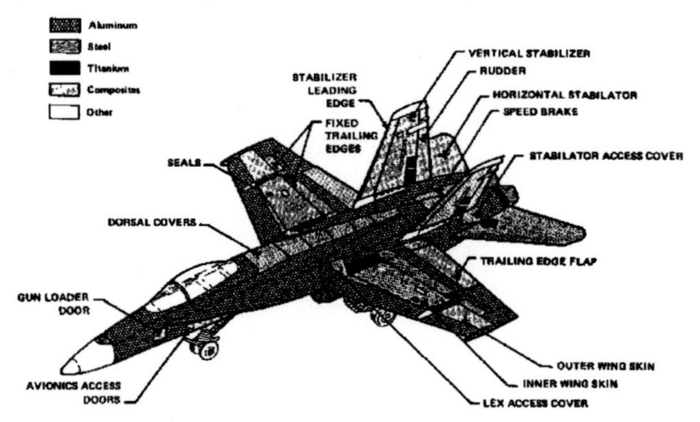

Composite distribution on a F/A-18. Its surface is 50% carbon fiber that could benefit from self-repair.

There are several approaches we tried for self-repair of polymers composites:

- Self-repair of barely visible damage in airplane components
- Self-repair in infrastructure such as walls
- Self-repair of pressurized systems

The system can contain the means of repair which is initiated by the same external influences that cause the degradation. A self-repairing system could provide "fail safe" design, rather than requiring redundancies that re-route functionality through additional extra layers. In other words, any overdesign for failsafe functioning can be reduced and a 15-30% weight savings and attendant fuel savings can be expected.

A self-repairing structure can be considered to recover functionality in the form of material characteristics, such as strength or modulus, or overall structural integrity. Physical and chemical degradation of materials, such as wear and corrosion limit the service life of the structure. Structural integrity may also be compromised due to unanticipated events, such as impacts. A structure that self-repairs would retain integrity without intervention, increasing survivability and potentially reducing maintenance downtime and costs for replacement and repair. Catastrophic failure of the composite structure will be prevented by preserving the ultimate strength above the allowable or operation load. Integrity of the composite structure will be preserved by retaining an adequate margin of safety for operation despite damage from impact. Self-repair technology will enable a passive application to a structure, which is reactive to the damage initiation impact event, for which structural integrity would be maintained without intervention. While micro-scale cracks are hard to detect unless they have developed to macroscopic scale flaws, we have used non-destructive evaluation techniques that have limited ability to detect micro-cracks. Any damage in the old method is normally repaired in the field by hand, but not all of the original strength is restored.

SELF-REPAIR OF BARELY VISIBLE DAMAGE IN AIRPLANES

The solution for airplane components is to develop polymer composites with unique toughness and strength by self-repair which occurs at material interfaces and at damaged areas by phasing out manual repair and instead utilizing release of repair chemicals from within the composite itself. The hollow fibers are embedded in the matrix, and the chemicals they carry are released wherever and whenever cracking or other matrix damage occurs. The repair chemical flows into the crack and crack faces are re-bonded. The in-air repairs take seconds.

The PHASE I SBIR done for Wright Patterson Air Force Base on self-repair in impacted composites processed by lay-up and autoclaving at 300-350F was successful, yielding restoration of flexural modulus of the non-impacted controls of 94% in fiberglass laminates for epoxy repair chemicals and 74-88% in graphite laminates for epoxy and derakane repair chemicals. Further, repair samples with twice the number of tubes and therefore more repair capacity have restored over 300% compared to the modulus in flexure of the control, which was impacted and an estimated 40% over the non-impacted control.

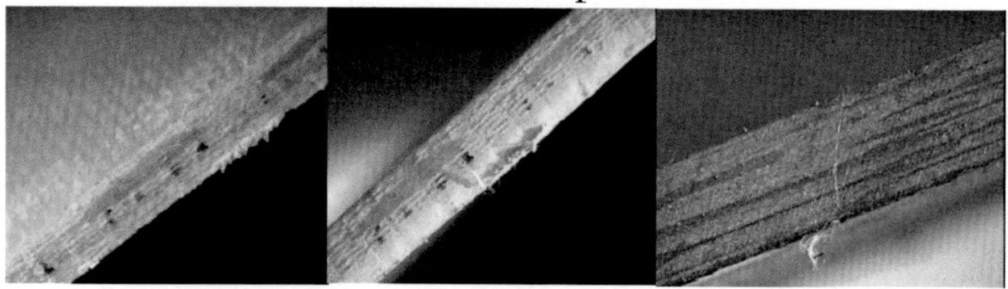

Left, areas of solid yellow adhesive can be seen in the areas of delamination, center, some yellow adhesive repair material appears in ply layers under the area of major delamination, and right, side view of repaired graphite epoxy shows repair in many layers of the composite.

NPD, Inc. has extensive experience in self repair and sensing of laminated composite materials. This prior work has included: repair using various types of stiff or flexible glass encapsulators, development of repair chemicals that can be processed under typical composite cure temperature/time profiles (250°F and 350°F and pressure), validation of repair durability under a range of loading conditions, and self-sensing. The figures which follow summarize key findings:

1. Repair chemicals survive processing heat in graphite laminates.
2. The system can re-repair
3. Repairs can be completed in less than one minute.
4. The repair chemical can flow into adjacent layers of the laminate
5. Impact damage is repaired, and original strength is restored (flexure testing: compression after impact testing (CAI); in-plane shear testing) Repair chemical re-bonds debonded fibers to the matrix

6. Self-sensing is done using eddy currents and color change shows reaction
7. Use of continuous, bendable, cuttable, coatable repair fibers in a laminate
8. Successful fiber insertion into composite processing

After the fiberglass samples were cut open along the repair fiber, it was observed that the area of delamination was filled with a hard, yellow looking adhesive pool that was solid. Furthermore, the repaired delaminations were proximate but not directly under the points of impact. The repair area is lens shaped. The yellow adhesive also flowed below this area to other ply layers of the laminate. The delamination area away from the repair fiber exhibited none of these phenomena. The pocket of resin was only associated with impacted points, which led to the conclusion that they are repair patches.

Left, repair chemicals survive heating in graphite laminates at 350F for two hours after one hour at 250°F. Middle, cut glass repair fibers after processing in the autoclave exhibiting liquid inside and right, when the laminations were pulled apart, the repaired mirror images on plies were revealed as repairs

Epoxy laminates exhibited regions of cured resin where delamination occurred next to the impact area. Both fiberglass and graphite laminates exhibited the presence of repair chemical in various layers despite having only one tube in one laminate layer without manual assistance. The examination cleaved samples entailed prying apart by a special insertion method to reveal the repair tubes. Graphite samples were made to be separated in the midplane to verify results in a visual non-quantified way.

Cleaved samples were very informative as they revealed the "look' of various types of damage and repair location in the laminates; for example, large tubes cut into the prepreg layers make the delaminations larger, smaller repair tubes have little effect One can see delamination's size, repaired area of delaminations and also note unusual effects such as the increased buckling of layers if repair fibers are cut into the prepreg.

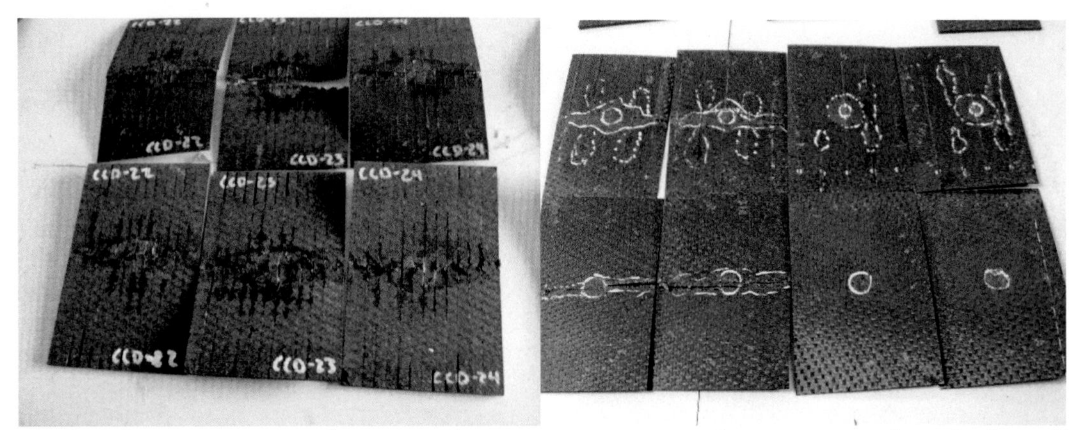

In this method of cleaving samples, the strips of prepreg cut to insert this excess number of tubes allowed the pulling apart of the laminate at the strip locations.

FLEXURE TEST OF REPAIR

Four-point bend tests were done on both fiberglass and graphite samples. The results showed a repair of 94% in the fiberglass and 85% in graphite laminates. Three types of controls were made: non-impacted samples no fibers, impacted sample no fibers, and impacted samples empty fibers. These samples were autoclaved using a vacuum bag at 250°F, then at 350°F for two hours. The samples were impacted and flexed to give quantified results.

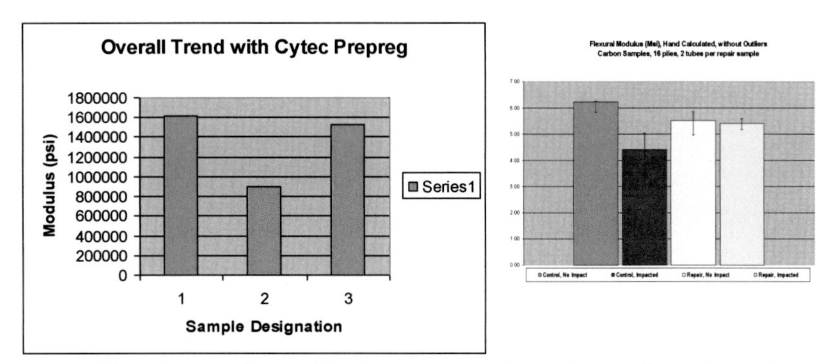

Repair in flexure tests on fiberglass and graphite laminates. Left, in the fiberglass laminate, the repaired samples at right repaired nearly all 94% of the damage. Right, in graphite samples the repaired sample in yellow restored the modulus to 85% of the controls in blue.

This SBIR Phase II effort was sponsored by the U S Air Force under a Small Business Innovative Research (SBIR) project. We focused on the aircraft structure which must carry the vehicle aerodynamic loads. Due to aircraft performance requirements (range, payload, speed, etc.), the structural weight is often minimized but structural integrity is assured through constraints of appropriate operating margins (stress, strain, buckling). When structural integrity is compromised, whether operating outside intended performance parameters or degradation of components, catastrophic failure can occur. Either redundant layers which add weight or self-repair can prevent such failures. Self-repair can maintain the desired low weight versus fail safe extra layers which add weight.

The objective of this effort was to demonstrate the feasibility of impact-initiated delivery of repair chemicals through hollow fiber architectures embedded within graphite fiber reinforced polymer matrix composites, representative of advanced aircraft component material systems. Self-repairing structures through coupon and elements were demonstrated and evaluated Barely Visible Impact Damage (BVID) was imparted to test coupons through height-controlled drop of an impactor initiating breaks in the hollow fibers, thereby inducing repair chemicals to flow into composite damage voids due to delamination and cracking. Effectiveness of self-repair capability was determined through comparison of measured ultimate strength of a standard set of pristine samples with specimens that were impacted, and specimens that were impacted and subsequently repaired.

Testing has been performed to ASTM Standards for flexure and compression loading, as well as fracture toughness. Shear loading was accomplished as an element test. The drop-tower impactor weight remained constant, 20 lbs., with change of height to reflect increased energy impacts. This experimental effort was conducted to provide quantitative, feasible proof of structure that repairs itself from impact damage.

Studies were also done on samples with variable parameters, a non-standard set of samples that are nevertheless tested in the same ways, through comparison of a standard set of pristine samples with specimens that were impacted, and specimens that were impacted and subsequently repaired. Both human and natural environmental hazards (tool drop, foreign object damage, hail) could produce Barely Visible Impact Damage (BVID) in the composite laminate, reducing the ultimate strength of the panel. The investigations have shown that significant self-repair could retain sufficient ultimate strength to maintain safety factor and margin of safety for operation, affecting survivability and sustainability of air vehicles. The goal is to repair any damage that has occurred in the matrix especially delaminations as well as cracks and voids.

In compression and shear, the adhesive and strength properties of the resin matrix are very important. The matrix must maintain the fibers as straight columns and prevent buckling, along with transferring the load across the fibers. Since only repair damage that has occurred to the matrix is attempted, tests were performed that isolated the strength the matrix adds to the material. These properties can be seen when compressive and shear loads are applied to the material. The main compressive test that was used is the Boeing Compression After Impact, BSS-7260 Test. This test involves impacting a specimen at a controlled energy level and then performing a compressive strength test, as shown in the picture, with three sides of the specimen clamped, and a compressive load applied to the fourth side. Delamination will cause a premature buckling at the delaminating interfaces.

COMPRESSION AFTER IMPACT TEST

Compression after Impact (CAI) tests were done on graphite samples to assess the repair of the matrix resin. The results showed a repair of 87% in graphite laminates. Three types of controls were made: non-impacted samples without fibers in blue, impacted samples without fibers in red, and impacted samples with dummy repair chemical and fibers in blue. Experimental repair samples in yellow were made with adhesives in fibers. These samples were autoclaved using a vacuum bag at temperature of 250°F for one hour and 350°F for two hours. The samples were impacted and then compressed to give quantified results.

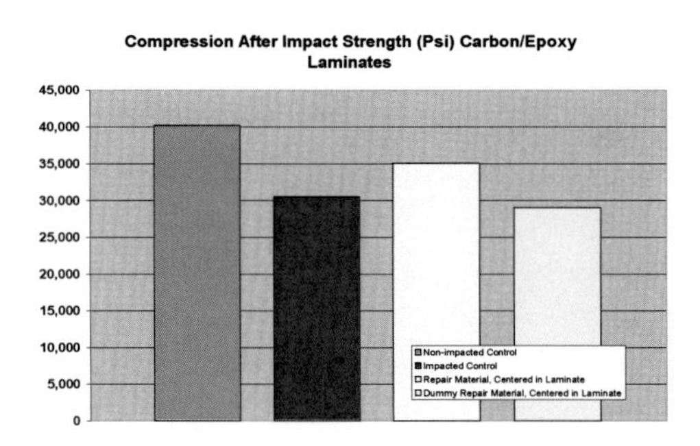

Compression after Impact (CAI) tests were done on graphite samples to assess the repair of the matrix resin. Three types of controls were made: non-impacted no fibers, blue, impacted no fibers, red, impacted dummy repair chemical in fibers, blue. The repaired samples, yellow bar, restored 87% of non -impacted controls, blue bar.

At left is a Compression After Impact sample in the fixture and at right is the sample after impact and compression loading. The failure occurred at the exact place impact occurred.

ELEMENT TESTING: IN-PLANE SHEAR TESTS

The test know as in-plane Shear BMS 4-23 specifies the procedure for determining the in-plane shear stress of a composite laminate. In this test, the laminate is loaded into a picture frame fixture and is tested under bi-axial forces to determine the in-plane panel shear strength. The laminate is constructed as a 9.5-inch by 9.5-inch laminate and placed in the window frame diamond fixture. However, doublers need to be adhered to the laminate, one on each side. The crosshead movement creates a tensile force on the top and bottom corners of the diamond shape while the side corners move in, creating a compression force. Testing occurs at a rate of 0.05 inches per minute and continues until failure. These need to fail at the same place, that is at the centerline where impacted for the data to be useable. It was critical to use enough impact force to generate such consistency; in these large shear test samples, the impact load that generated consistent buckling was 500 inch-pounds. The repair sample in black had a peak stress that is 82 % restoration of peak shear stress strength when compared to the non-impacted control in white. It is a 36 % improvement over sample with tubes only, in stripes; that is the contribution of the repair chemical.

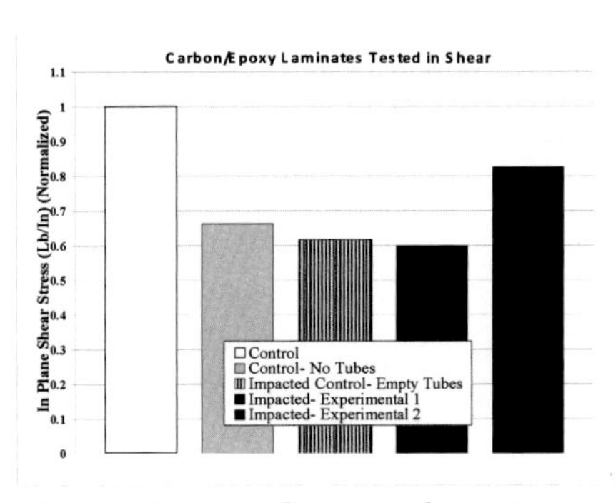

Left, chart of the data points to date on shear impacted at 25 inches with a 20 pound weight for a total 500 inch pounds. (Right) typical sample for shear testing.

SELF-REPAIR IN INFRASTRUCTURE SUCH AS WALLS: SELF-REPAIRING AND SELF-SENSING WALL PANEL COMPOSITES

This work addressed a U.S. Army Challenge for multiple-impact superior reusable panels and stanchion wall systems. The field test-ready self-repairing and self-sensing composite protective system met the Army challenge for improved force protection for combat outposts and patrol bases. Technology capability demonstration with aligned manufacturing readiness was ensured through this effort's collaborative work with two composite manufacturing companies.

Left, Isometric view of Elastic 3D Solid with 24 elements thru-thickness at maximum deflection expected upon blast with 25 pounds of C4 at 20 feet, Center, a thrice blasted panel shows the pattern of similar deflection in the refringent pattern Right, high speed video pictures showed the shock wave as a wave then the fireball blast that deflected the panels backwards and then the vacuum that pulled panels forward.

All panels (controls and self-repairing) remained intact after a series of 3 full blast tests with 25 pounds of C4 at 20 feet. Data including blast pressures, wrap around pressures, high-speed photography and simulated grenade and ballistic testing revealed that the self-repairing panels retained their ability to stop impacts after the 3 blasts were completed, while control panels did not stop these. Three nearly week-long blast sessions were held and each session the panels were blasted three times with 25 pounds of C4 at 20 feet and then were tested for impacts. In the move up the Manufacturing Readiness Levels designation, NPD Inc. made self-repairing panels at two companies.

Left, arena style blast set up of NPD panels, Center a velocity analysis on an aluminum only panel showed the velocities, Right, panels and wall stanchion structure after third consecutive blasts.

Left two, the self-repairing panel front view and witness plate from the back showing very few penetrations in panels that had had several large blasts prior, Right two, control panel front and witness plate showing many more hole penetrations in panels that had had several large blasts prior. The self-repairing panel stopped 95 % of the penetrations while the control stopped 54%.

Left two, making self-repairing panels using the hot press process, Right, the self-repair system inserted into the pultrusion process.

The most interesting finding was that the self-repair system is easily inserted into the both the pultrusion and the hot press systems.

SELF-REPAIR OF PRESSURIZED SYSTEMS: BICYCLE TIRES

One of the most effective uses for self-repair is utilization in pressurized systems because the pressure helps to push the repair chemical into the damage site. For the repair chemical to stay in the damage site and repair rather than be pushed out by the pressure, the repair chemical must be designed appropriately for the pressure and volume of air of the system. In each of the five examples of self-repair in pressurized systems, the viscosity, chemical makeup, and volume of the repair chemicals used is different for each system.

In initial bicycle tire repair, the pressure was about 60 psi and so the chemical used was a fast reactor and thin. In initial automotive tire tests, the pressure was about 35 psi so the reaction rate of the chemical was slower than bike tires and chemical was more viscous. In air beam tests the pressure was 60 psi (more air volume than bike tires so that called for a different chemical. Other examples are NASA self-contained habitats which have 15 psi inside and a vacuum outside with large volume of air or gas and high-pressure pipes which can have pounds of force per square inch of 700 to 1400. The solution for these is to use a very fast reacting repair chemical and then assess what hoop strength is created after impacts and so predict the pipe pressure on which it will self-repair.

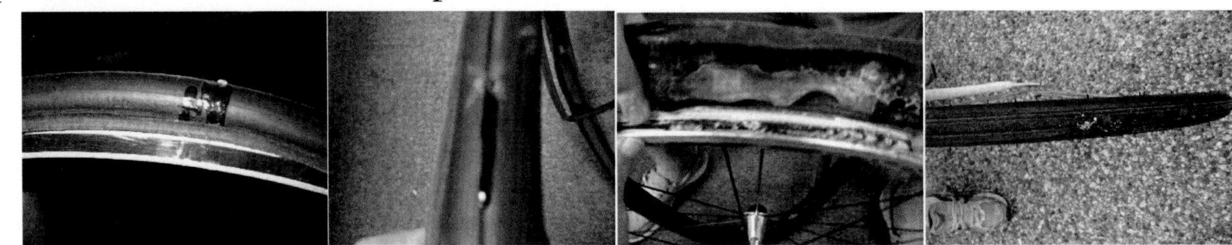

Left, the tire with repair chemical coming out of the hole made by an icepick. Next picture, the repair chemical is flowing out due to a squeeze put on the tire. Next picture, the assembly pulled apart after repair to reveal the release of the repair chemical onto the inner surface of the tire. Next, the inner tube is bonded completely to the tire. Right, The repaired tire after puncture, repair and riding on gravel.

In labs scale tests on bicycle tire self-repair, NPD, Inc. used repair chemical which was placed around the whole inner tube and adhered between the inner tube and the inside part of the tire. After the tire was mounted, an ice pick was used to puncture the tire and inner tube, allowing the repair chemical to flow out and repair the damage to the tire and inner tube. The pressure was about 60 psi, so the chemical used was a thin, fast reactor.

In lab tests using bicycle tires with scalable results, NPD, Inc. found the tires could repair within a period of one second and that there was minimal loss of air pressure in the first two hours. There was minimal additional loss of air pressure over a week's time with the ice pick or nail being able to be left in or removed with both repairs being successful. This showed repair would occur with a heavy rider.

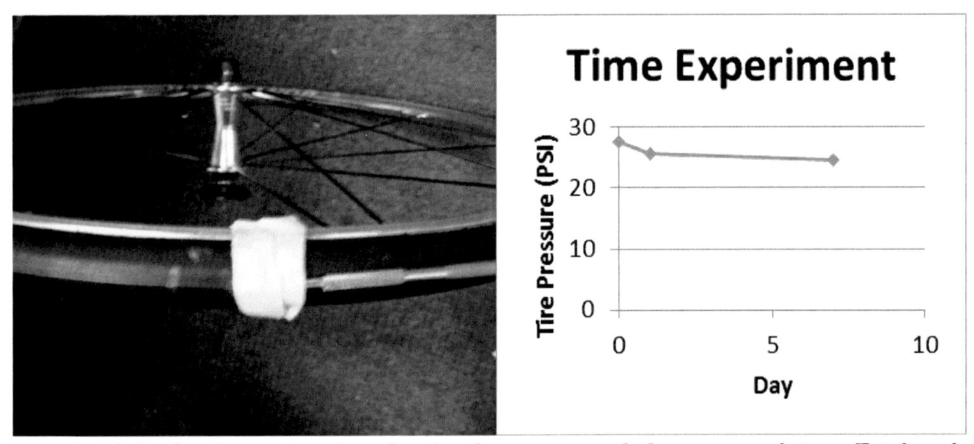

Left, an example of the hook puncturing the tire for an extended amount of time. Right, the tire pressure remained nearly the same after repair for a period of one week.

Successful trials in this experiment meant a successful repair occurred without pressure loss. Two trials were successfully tested, beginning with a tire pressure of 30.5 psi and 27.5 psi respectively, and ending with a tire pressure of 27.5 psi and 24.5 psi respectively. Further, a successful trial meant the tire would repair after a puncture lasting over an extended amount of time (2 hours). This experiment simulates a foreign object being lodged in the system and then dislodging, therefore creating a puncture in the innertube. After removing the object (a hook), which had been in the tire system for two hours, a normal reaction between the adhesive and tire was observed. No tire pressure was lost over the two-hour period, and no air was felt or heard leaving the assembly during repair.

In a further test, after a successful puncture and repair, the tire was observed for a week, checking pressure throughout to ensure no pressure loss. After one week, the tire pressure was not significantly different that the initial measured pressure. The beginning tire pressure was 27.5 psi, and the final tire pressure was 24.5 psi.

Next, an experiment was done to determine if the self-repair could be successfully conducted on a full-scale bike with a rider. A secondary purpose was to determine if particles from the road would affect repair or the ability to ride the bike after repair. After puncturing the tire on a series of nails in a board, the bike was ridden around on the street to see if particulates would attach to the tire. The tire was repaired during testing, and there was some dirt and small particulates on the repaired area, though nothing significant enough to affect the ability to ride the bicycle. The beginning tire pressure was 47.5 psi, after testing pressure was at 40.5 psi. After one week, the pressure was 39 psi. This shows that the repair was successful.

In final tests, the bicycle tire was successfully punctured, ridden, and repaired five times in various locations. The tire fully repaired by no loss of tire pressure over time. The fact that the tests ranged over a long period of time illustrates this concept well. The insert successfully repaired the bike's puncture 5 times before finally failing to self-heal. The tests ranged over a period of four months, showing a reasonably long lifespan for such a product. After tire deflation, the repair chemical was taken out and some of the self-repair chemical was still unreacted, showing that more punctures could be repaired.

SELF-REPAIRING AUTO AND TRUCK TIRES

In automotive tires subjected to impacts with pressure of about 35 psi, the reaction rate of the chemical was slower than bike tires and chemical was more viscous. Assuring the safety of tires that carry people is critical, and heavy runflat tires are the accepted means to do this. I have developed a much lighter alternative way of ensuring the viability of punctured tires and have tested the concept on small trucks. I have shown that tires can have a layer of releasing repair chemical that react in 1 second, thus repairing the tire rubber and maintaining the air pressures. In a repeated damage event, the penetration of debris will not be successful because the prior damage to the tire would be repaired.

The self-repair chemical has been formulated to remain reactive under the heat of extreme tire use. In the tested 3-ton vehicle, time for the solution to take effect equals lost air pressure

so a fast reacting and quick flowing liquid chemical may be preferred. If a runflat is used, the rubber tire will be ruined; it is like running on the rims in a car after a flat tire. It is difficult to change a runflat tire. With the self-repairing tire, if it is punctured some of the air escapes but the repair
chemical is released into the damage site and repairs it, restoring air tightness. Thus, the rubber of the tire is saved and pressure is maintained.

Left, puncturing a truck tire. Right, inside the repaired tire.

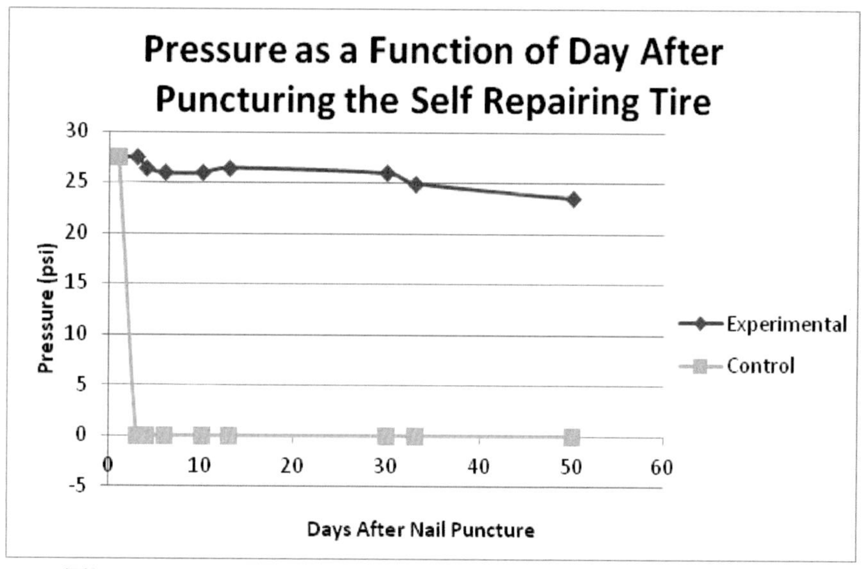

The tire pressure remained nearly the same after repair for a period of one day.

The requirements for self-repairing tires are: the system must repair damage, the full reaction must be completed in a second after impact, repair chemicals flow and react easily inside tire material, the repair system can re-repair many times, more than two times, the tire must be repaired to 95% of virgin load capacity in flexure, the tire holes must be repaired to 95% of non-damaged state, there can be little to no loss of air pressure after damage and repair, the vehicle load must be supported during damage, the repair chemical must survive the heat and pressure of road use the chemical survives wet heating, freezing as unreacted and as reacted, the repair chemicals are durable in shelf life of 10 years, simulated fatigue.

In this test the tire without runflat was mounted on a wheel was inflated to its pressure limit and loaded. The tire was then punctured and pressure loss and time to deflation was measured. Tires for each testing regime consisted of controls with no repair system and experimental samples with repair chemical only in varying delivery systems The tires were punctured several times to assess the self-repair function. After all the testing, data and change in hole size information was obtained and the tire panels were visually inspected.

Durability tests were done on the tires after they had been punctured and repaired with low amounts of tire pressure is lost. Over the course of 50 days, 4 psi was lost in the tire and further tests have shown that the majority of that is from successively testing the pressure of the tire. Every time that the tire gets its pressure tested, it loses some air. That was the main loss of tire pressure over the 50 days.

The self-repair chemical is a polymer combination of flexible, strong polymer with several additives to affect additional rupture arresting, fast reaction capabilities and a transformable color sensor. The chemicals are contained in small rupturable tubes made of polymer, metal and fiberglass. Upon impact, the tubes break and the chemical is pushed out into the new damage voids (such as holes and delamination) and sets up when it touches the damage site. This is achieved by the complex, material-specific formula reaction, thus repairing that material damage and filling any breaches. This process provides visual verification by indication that the chemical has reacted with the damage surfaces by producing an indicator color.

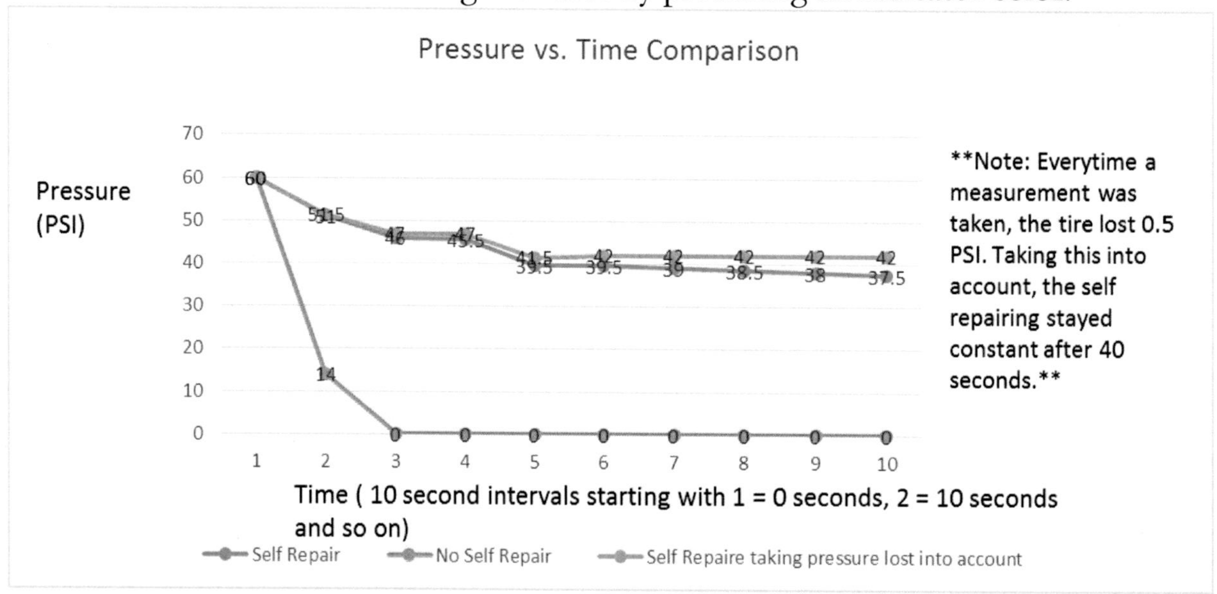

Tests on truck tires comparing time and pressure for self repair and controls

SELF-REPAIRING AIR BEAMS

In air beams that survive impacts, the pressure was 60 psi. They had more air volume than bike tires so a different chemical for quick repair but slower evacuation was required.

This technology integrates novel self-repair technology into air beams by providing improved multiple impact survivability over controls and by mitigating the release of pressure upon puncture associated with air beam technology. This repair achieves near instant re-stabilization of damaged air beam material and pressure integrity while maintaining virgin material structural integrity over multiple impacts. Embedded visual damage and regeneration assessment, achieved with color, aids in damage and regeneration effectiveness assessment. This optimizes life cycle and extends virgin equivalent material protection integrity.

Air beams tested at NPD. Far right, an air beam from HDT holding a car.

The overall objective with this effort was to demonstrate, in prototype form, the feasibility of novel self-healing technology in air-beams. The solution provides vastly improved impact multi event/ structure protection, visible repair inspection (activated by a color change), offering unprecedented repair and damage assessment for fabric shelters.

SELF-REPAIR OF PRESSURIZED HABITATS FOR SPACE USAGE

While testing NASA composite habitats, we demonstrated repaired impacts under vacuum which has 15 psi inside and a vacuum outside. These also had large volumes of air and hence required a chemical similar to the truck tires. We looked for improved resiliency and improved survival of space infrastructure when impacted by micrometeorites or space debris. Gaining resiliency and reliability benefits over the current status and covering any emerging damage threats while maintaining infrastructure integrity over time while and maintaining maneuverability during full spectrum operations via multiple impacts (space debris and micrometeorites) were our goals. The problem is space debris impacting manned and unmanned space infrastructure.

At the low space altitudes at which the International Space Station orbits, for example, there is a variety of space debris. Large objects could destroy the station but are less of a threat as their orbits can be predicted. Objects too small to be detected by optical and radar instruments from approximately 1 cm down to microscopic size number in the trillions. Despite their small size, some of these objects are a threat because of their "kinetic energy and direction in relation to the station." [1] The space station goes to great lengths to shield inhabited pressurized spaces from debris to avoid failure of a component.

Left, a 7 gram object (shown in center) shot at 7 km/s (23,000 ft/sec) (the orbital velocity of the ISS) made this 15 cm (5 7/8 in) crater in a solid block of aluminum. Next, the International Space Station has capacity for materials experiments both inside and out in space. Right two, proposed space experiments for self-repairing pressurized infrastructure for testing on space station.

Specifically, this technology integrates novel regenerative and self-repairing technology into pressurized wall armor formulations, or into air pressurized systems providing improved multiple-impact survivability over current structures by eliminating the pressure loss due to particle penetration. This method also mitigates damage and delamination associated with composite space structure technology. The regeneration achieves instant re-stabilization of damaged walls and maintains virgin pressure and impact protection over multiple impacts. Embedded visual damage (regeneration assessment), achieved with added color, aids damage and effectiveness assessments of the structure.

Left, the NPD test series, all of the regenerating/ repairing samples showed successful resistance to depressurization with no pressure loss (less than 4% pressure loss due to pressure measuring), Next, in accelerated testing the regenerating samples showed successful resistance to depressurization with no pressure loss (less than 4% pressure loss also), Next, the regenerative composite wall sample is placed over a vacuum pot and then punctured with a 0.075" metal punch tip that made a 0.155" diameter hole. Next, the vacuum gauge showed that the pressure of – 27 inches of mercury were maintained after the puncture and subsequent removal of the metal punch Right, there was no evidence of repair material pulled from sample into vacuum pot.

TESTING RESULTS

NPD, Inc. has developed regenerative technology specifically designed for application to pressurized walls in the vacuum of space subjected to high speed impacts of micrometeorites and space debris. With more than 300 prototype experiments fabricated, the optimized regenerating wall technology for pressurized habitats was developed. The figures summarize the results for the pressurized habitat walls.

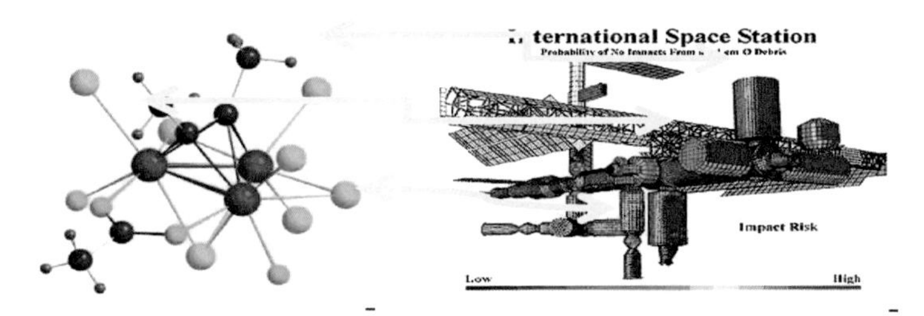

Diagram of the self-repair chemical applied to the space station areas needing it.

SELF-REPAIRING HIGH-PRESSURE PIPES AND PIPELINES

In self-repair of moderately high-pressure pipes, only hoop strength was measured as a way of assessing what size pipes under a specified pressure could be repaired. The pressures can be so high and the pipes can be so large that no fast-reacting repair chemical would work. For example, oil pipelines with a 20-foot diameter can have pressures of pounds force per square inch of 700 to 1400 (which is extremely high). Smaller pipes can be self-repaired, however.

The objective was to find the apparent hoop strength of self-repairing composite pipes with chemicals added to the pipes. Once apparent hoop strengths are found it can determine the effect of adding self-repair to the pipe and what pressure it might be able to withstand.

Pipes are made and designed to carry different types of fluids. Composites pipes are beneficial due to the low strength-to-density ratios. Filament winding is used to create composite pipes and different materials can be used. Rovings are most commonly used in filament winding and are wrapped onto the mandrel in different orientations. The two main wrapping styles are helical and hoop windings. In helical winding, the roving is applied to the mandrel at an angle less than 90°, whereas, a hoop winding applies the roving at 90° to the mandrel. These two winding styles affect the properties of the finished pipe. A way to determine how strong a composite pipe is to run an apparent hoop strength test on hoop wound samples which is what we did. This test determines the comparative apparent tensile strength of each sample. To do this test, a testing fixture shown in ASTM D2290-12 was made to pull the samples apart.

The reinforcing material was E-glass roving. The matrix used was a resin system. All of the tubes were fabricated on a filament winder and spun until gelled so the resin was evenly distributed throughout the sample. The tubes had 8 layers of fiberglass and self-repair was added to some samples to determine the effect it had on the apparent hoop strength. After tubes were cured, the procedure below was used to prepare the samples for testing.

A self-repairing pipeline could limit environmental damage. CC Image Kevin Abbott via freeimages.com

TESTING FOR HOOP STRENGTH

The pipes were wound with 8 layers of hoop winding, some with self-repair and some as controls. Then, the pipes were cut into sample specimens according to the ASTM standard ASTM D2290-12. After that, a 180-inch pounds of weight from a drop tower was dropped onto the sample, the sample was put into test fixture, exerting pressure from inside. Finally, pressure was applied and measured.

When self-repair was added to the pipe, it had an overall effect of increasing the average apparent hoop strength slightly. This was due to the self-repair flowing into the voids caused by impact. Once set up, the composite sample could transfer the load through the repair and matrix so the fibers could take more load in the hoop strength test.

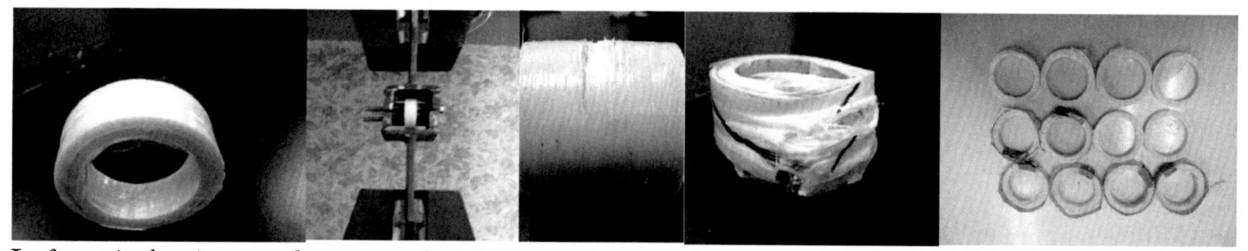

Left, typical unimpacted test sample, Next, test fixture shown in ASTM D2290-12 with sample ready to test. Next, tested and broken self-repair roving samples. Right, set of pipe samples

One can also reverse the equation for pressurized systems and get hoop strength for purposes of comparison. The repair chemical must be designed appropriately for the particular pressure and volume of air of the system. In each of the five examples of self-repair in pressurized systems discussed, the viscosity, chemical makeup and volume of the repair chemicals used is different for each system based on the hoop strength needed AND for the force and size of damage to which it is subjected or an acceptable factor of safety.

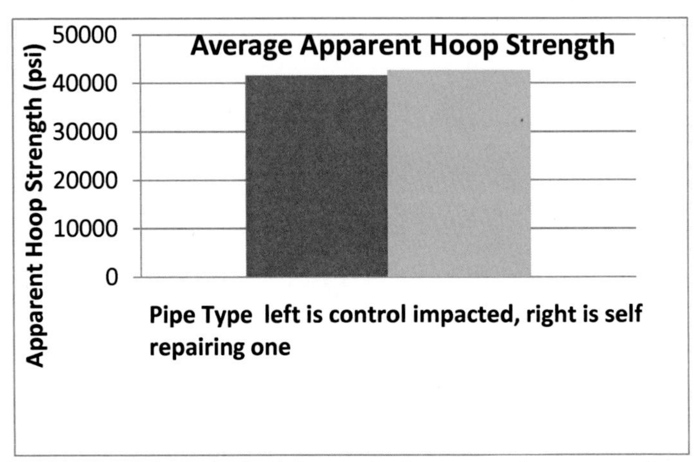

Apparent hoop strength of roving pipes (psi); impacted pipe controls compared to impacted self-repairing pipes. In self-repairing pipes, there was an increase in the average apparent hoop strength, slightly.

CHAPTER 6:

SELF-SENSING IN CONCRETE AND POLYMER COMPOSITES

Self-sensing and self-repair go hand in hand. Damage can go undetected in polymers as well as in concrete; having some idea of the damage and the repair accomplished allows manufacturers and users to be more confident in their use of lighter weight and more efficient materials.

Polymer composites are especially prone to undetected damage that can be catastrophic. Verification of damage state and self-repair status would eliminate the uncertainty about safety and reliability. Concrete has similar drawbacks but also is opaque and therefore difficult to "see inside" to assess damage. There are numerous approaches I developed for self-sensing and have organized them into two categories: self-sensing in opaque materials and self-sensing in polymer composites.

It is difficult to observe damage inside opaque materials like concrete in a dependable way, yet the damage can be catastrophic. A goal is to provide information on the location and size of the internal damage without using new, disturbing probes. I aim to record the self-repair fibers status: their breakage, location and number of breaks, and size of fluid loss for sensing damage size and location. This adds confidence to continued use of structures and their repair.

FIBER OPTICS

The goal of this research was to develop liquid core optical fibers for the detection (and self-repair) of cracking wherever and whenever damage was generated in cement, polymer, or other semi-opaque brittle materials. The presence of adhesive liquid-filled fibers inside an opaque matrix such as cement suggested the possibility of using these hollow, liquid-filled repair fibers for the nondestructive internal evaluation of the material as well as for repairing cracks. Shining a laser beam or bright light through glass tubes filled with adhesives as a means of analyzing the behavior of adhesive systems inside concrete samples was one example of assessment.

When using a laser, it was shone through a glass tube at a point where a photosensitive optic diode detected the power of the resultant beam, and translated that information to a voltmeter.

Glass tubes with various amounts of adhesives allowed variable amount of laser light through them to the detector. The amount of laser light passed through the tubes was used for evaluating how much adhesive remained inside the tube, though the method was an indirect way of measuring volume of cracks. By observing the laser beam diffraction patterns that exit the other side of the filled glass tube, a base was constructed for the laser with aligned projection screens that were used to standardize recordings of the laser diffraction patterns. By projecting diffracted laser light on a screen at a standard distance from the end of the capillary glass tube, it was observed that different light patterns appeared for different adhesive configurations or patterns inside the capillary. Whena capillary was filled with an adhesive end to end, a strong outer ring (projected by the glass) and a fuzzy blob of light within the ring (projected by the adhesive) was observed. When a capillary was filled from one end to the middle and the light was projected through the full end, two solid concentric rings on the screen were formed, a large one for the glass, and a smaller one for the meniscus of adhesive in the middle of the capillary. These light patterns correlated with the readings of voltmeter.

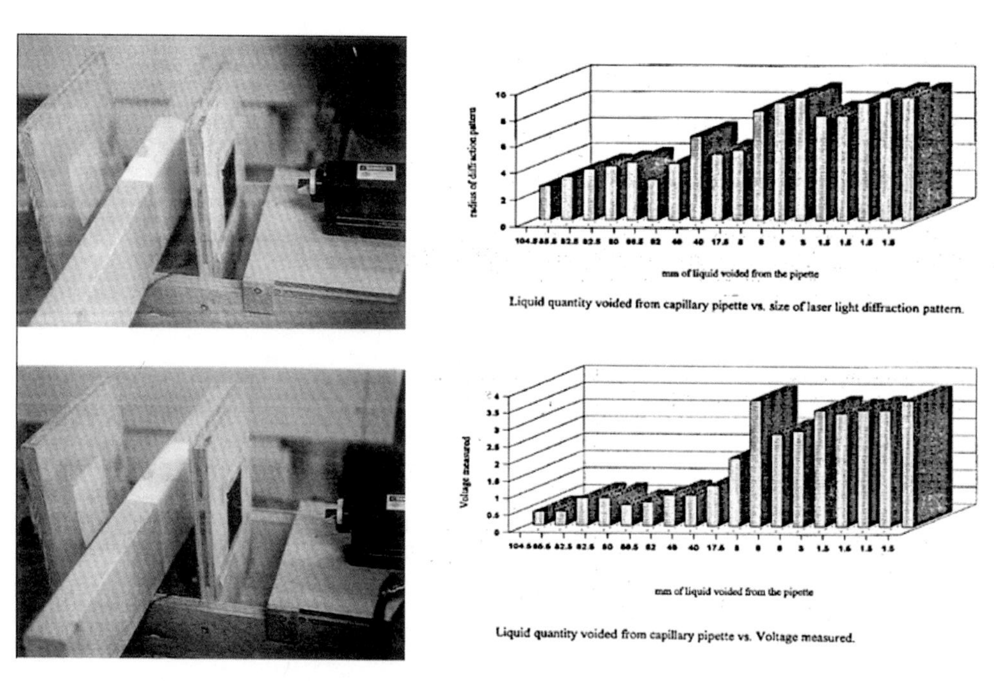

By projecting diffracted laser light on a screen at a standard distance from the end of the capillary, it is observed that different light patterns appear for different adhesive configurations, or patterns inside the capillary. In the bar graphs, the results show a linear relationship between adhesive quantities and the resultant laser light power as measured by a diode in above graph and graph below that shows adhesive quantites and the diffraction pattern size as represented in bar graph.

The laser light diffraction patterns and beam strength were directly related to the amount of adhesive in the pipettes. The results showed a linear relationship between adhesive quantities, the resultant laser light power, and diffraction pattern size (see figures). This is a new idea addressed an old problem in urgent need of a solution: crack location and determination of crack volume.

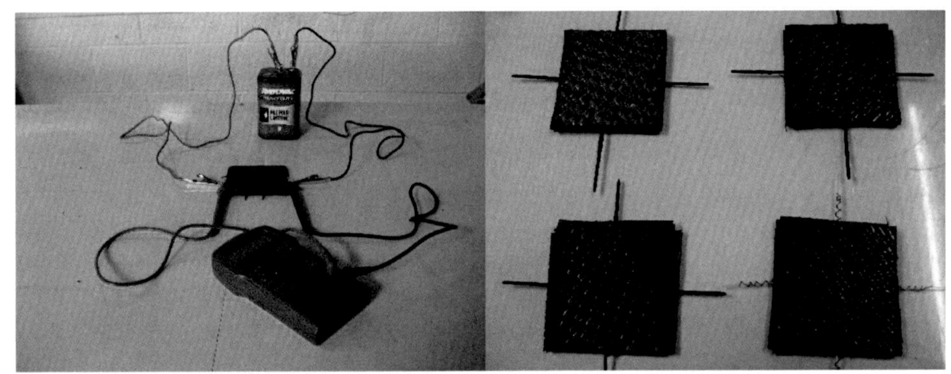

Left, test set up for measuring conductivity. Right, samples in which coated tubes project outside and can be connected to the conductivity measuring apparatus.

The repair tubes were used as fiber optical probes to assess the location and volume of damage. The remaining repair chemical could project a pattern which could reveal location of loss of repair chemical and volume lost. An independent verification was done using an OTDR. A light is passed into the glass fiber. More specifically "an OTDR transmits an optical pulse through an optical fiber and measures the fraction of light that is reflected back… By comparing the amount of light scattered back at different times, the OTDR can determine losses. The OTDR displays the optical signal as a function of length" [12]. Therefore, one can pinpoint the point of loss or a fiber crack.

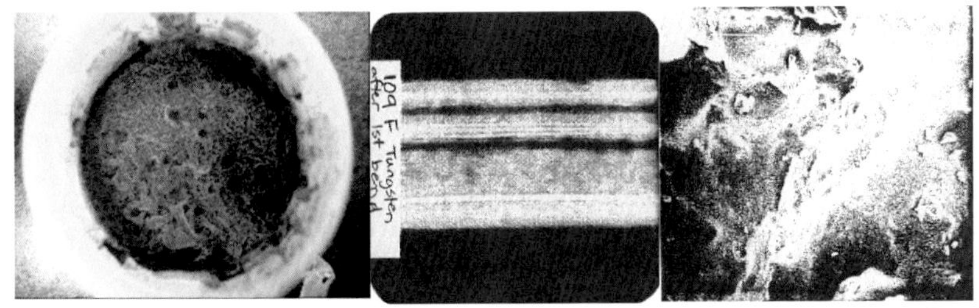

Left, powdered carbon (also powdered nickel) was mixed with the repair chemical, Center, powdered tungsten was mixed with the repair chemical. It appears as small white specs in the x-ray photograph and as the light material is the SEM photo of a flow of repair chemical. Right, nanotubes mixed with the repair chemical

The repair and sensing fibers and chemicals system can be used to sense (in three dimensions) the location of the damage event, detection of damage size, detection of self-repairing, detection of amount of repair, and determination of residual strength of the component by using a unified fiber structure designed to function inside 3D composites. Repair fibers were placed in a grid network over several composite layers. The location, extent and damage (stress) level was sensed by measuring the optical reflectivity of repair fibers.

Polymer composites are comprised of two materials: the resin matrix and the fibers. Polymer composite structures vary depending upon the layup process utilized. Most generally they are composed of thin layers of prepreg which can be woven or unidirectional tapes or fiber rovings inserted into a wet resin as in pultrusion. The overall design of the layout of layers, each of which contains fiber and resin, is such that fiber orientation of the fiber layers are switched back and forth differing angle to make a strong overall composite. The layers are balanced sometimes symmetrically about the middle ply so that the composite does not bend. The figure is a drawing of prepreg ply layouts for two balanced composites. When it is made up of layers, the composite is called a laminate.

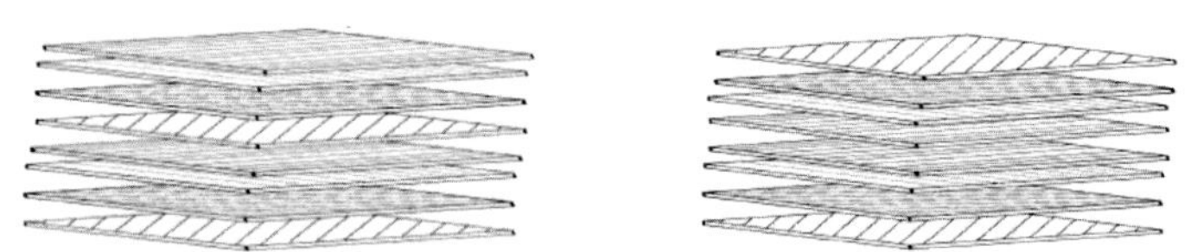

Left is a quasi-isotropic layup and right is a layup symmetrical about the middle.

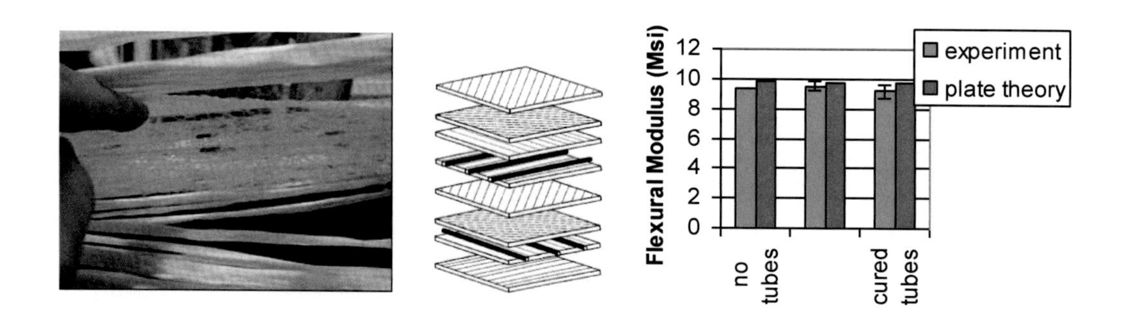

(Left) adhesive filled tubes are inserted by NPD into a pultrusion roving, (middle) is a problematic layout in which the repair/sensing fibers are laid out correctly with the overall 3d design of the laminate and (right) the data showing that putting repair fibers in the neutral axis do not affect properties

Hollow glass fibers with a transparent and liquid repair chemical inside are a simple fiber optic. When placed in a spatial pattern inside the laminate, the illuminated fibers create a light pattern that allows operators to record where a fiber has broken in a three-dimensional grid. Fibers may be overlapped (one above the other) to ensure proper mixing when both fibers are broken for proper release of the chemicals and to ascertain damage location in three dimensions. This design could entail alternating the two chemicals in the web, one in one direction, the other at right angles to it to ensure proper mixing and sensing. The insertion into weaving of the sensing and repair fibers can be done in the warp and woof. Specifically, the number of fibers involved that are spread out in the laminate gives information on the size of damage. The sum of all the broken fibers tells the extent of the damage as well as locations. The fibers in opaque laminates are surveilled from the ends.

Also, the repair/optical fibers can be telescoped inside each other. Damage will break both fibers, potentially causing mixing of both chemical parts in proper ratio. This double fiber is useful as a double fiber optic with the opportunity for dual mode information transmittal. In all these cases, the fibers can be examined for sensing of damage location and size.

Coated tubes can be sensed and determined if they are broken but also as with the fiber optics interrogated ones the layout of tubes in a 3d pattern can reveal aspects of damage. Self-sensing using eddy current scans on samples with metal-coated tubes clearly detected the metal-coated glass fibers in two directions to a depth of .21 inches. Cracked or broken metal coated fibers can be detected as a spatial change. Variations in the metal coating thickness, width and placement on the fibers can be seen. The sample specimens were slightly more than one-quarter of an inch thick and the metal coated fibers were placed at most one-quarter of an inch away from the surface. It can be observed that fibers that are further away are less sharply in focus. The depth of sensing is based on a rule of thumb: the sensor can easily sense to one half the depth of the coil. To go deeper, electrical output must be reduced, though it is harder to sense, hence the emphasis on highly conductive metals. Also, repair chemicals change color upon reaction, indicating that the repair occurred.

Left, any break in the hollow optical fibers emits laser light at the break site. Center, hollow glass fibers containing repair chemical laid out in a grid so that locations of break can be pinpointed. Right, nested optical fibers which contain repair chemicals.

When inserted into the composite architecture in 3 dimensions, the very layout of the sensing and repair fibers can be used to further the goals of repair and sensing. A series of liquid filled hollow metal coated glass fibers with chemicals inside can pinpoint damage in 3D when any are broken. Metal coating on repair fibers and conductive repair chemical are surveyed from the outside of the composite with eddy currents. This sensing can indicate in a timely manner whether there has been a damage event, the location and size of the damage in 3D and if the repair chemical has filled the damage.

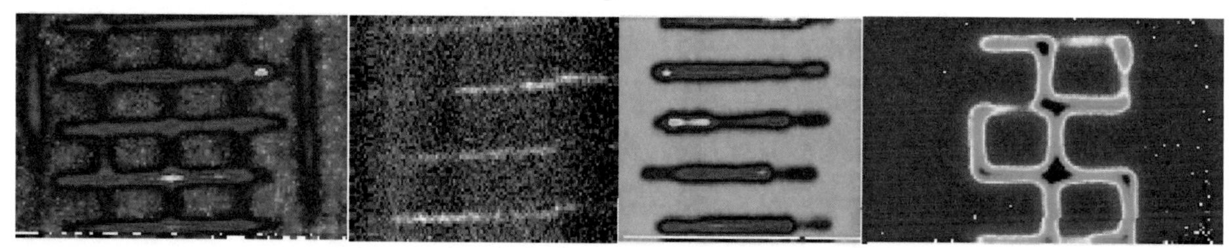

Left, eddy current sensing shows metal coated fibers at right angles to each other in a .25-inch-thick laminate in which the fibers are placed at a depth of and at .21 inches to the top of the fibers. Center, broken and discontinuous jogged fibers revealed by eddy current sensing. Center right, fibers show variations in thickness, width and placement of metal coatings by eddy current sensing. Right, a grid of using eddy currents. Repair tubes are coated with a metallic paint that carries a charge; the paint shows up in an eddy current scan which proves they are conductive. Earlier tests confirmed this too.

The NDI ultrasound tests clearly showed the repaired delamination areas in the middle of the repairing sample and some repair of delamination caused by the repair tubes in the laminate.

Ultrasound was most useful in comparing the stages of repair at various second intervals after repair is initiated. However, this use of ultrasound did not yield as much information as other inspection methods in general.

Left, the composite sample ultrasonically monitored before impact. Center, the same location in the same sample tested after impact in which the impact circle and detached fibers are seen as yellow. Right, even later after repair has taken effect, the horizontal fibers are reattached (no longer yellow) and the impact circle filled.

These ultrasound sample photos made at different times can indicate the progression of damage. A comparison between the scan without impact, left, scan of the original impact on the experimental sample, center, and of that sample some 45 minutes later in exactly the same place in the sample reveals the following:

1. The impacted area (yellow circle in the center figure) is somewhat repaired in a comparison of the original impact damage (left figure).
2. The delaminations caused by tubes insertion reads as two large horizontal swatches. These diminish after 45 minutes as the repair chemical re-bonds the tubes to the matrix.

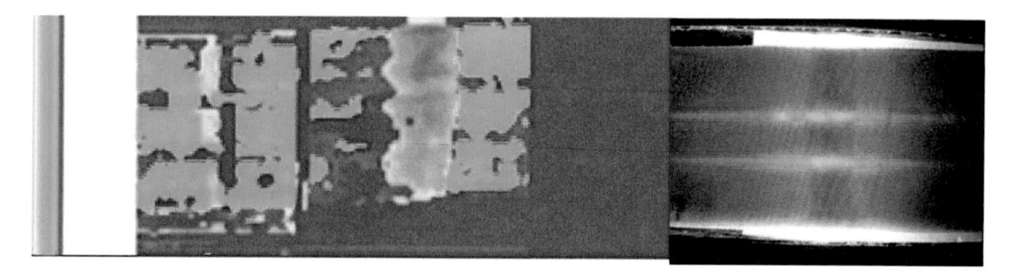

Left, the ultrasound record of testing of the sample pictured at the right and a control.This one colored visual of an ultrasound test shows a repaired sample on the left with no orange and right is the unrepaired sample that contained no repair chemical and so have a large damage area seen as orange. The samples had been sliced in the middle and flexed with a hammer at that location. Right, a photo of the repaired sample. In this photo the red dyed repair chemical can be seen to have voided the repair tubes in areas that appear lighter and it has gone into the damage zone in the center of the specimen.

Videos show that a laminate subjected to impact quickly makes delaminations and fiber debonds which are quickly filled with repair chemical. The path for the chemical to flow along is made by the damage itself. When that damage is filled, the flow stops.

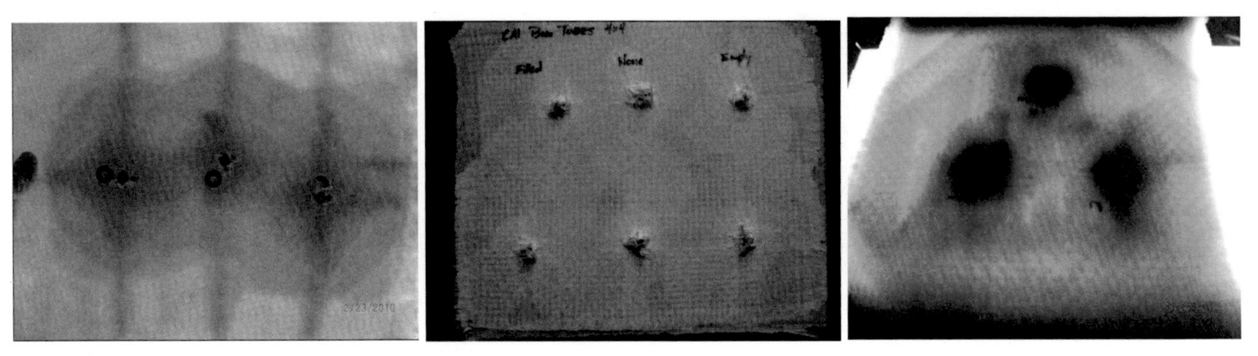

Left, 9 ply panel made from sc15 that has been impacted, in which you can see the delaminations on the back side and the flow of repair chemical in three locations where the tubes were filled with repair chemical. The yellow color indicates where the repair chemical has flowed and also that the chemical has reacted with the matrix. This panel was the designed as an open hole test. Center, an 18 ply panel made from sc15 that has been shot, in which you can see the delaminations on the back side and the flow of yellow repair chemical in the left side which is the only location where there were tubes filled with repair chemical. This panel was the designed to test in cai later. Right, a backlit 24 ply sample which was impacted and the yellow repair chemical can be seen to have flowed into and repaired the delaminations.

The actual repair by chemical reaction is indicated by a color change of the repair chemical. It is yellow when reacted with the matirx and red if it is exposed to ambient air such as in a fiber bust out into the air from a ballisitc impact.

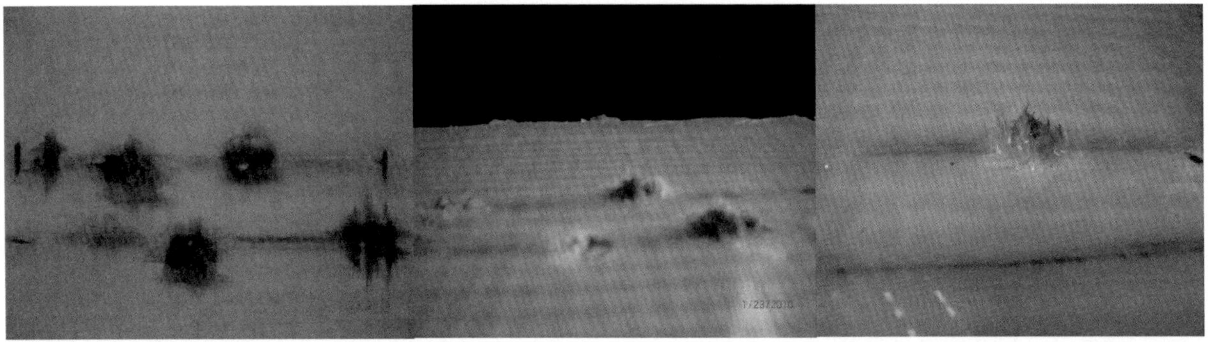

Left and center, at impact sites, the repair chemical turns yellow to indicate reaction is completed. Right, the yellow reacted repair chemical turns red as it saturates the broken fibers as it is exposed to the air. The broken fibers are reattached, merged, bonded back together in the burst out.

CHAPTER 7:
RECYCLING

The manufacture of cement generates 8% of the world's CO2. The goal of recycling was to extend the service life of concrete structures such as bridges without ruining more than was necessary so less cement needs to be produced. The biggest advantage of any type of recycling is the saved materials. In this example, pre-stressed concrete girders have huge amounts of explosive energy embodied in the pre-stressed tendons. The girders must be carefully demolished so that other parts of the concrete structures can remain useful. their demolition is very complex and dangerous.

Pre-stressed girders are the most difficult construction components for demolition because they have embodied energy. They tend to explode all around them as the pre-stressed tension in the steel tendons are released. Most bridges and many buildings use these types of girders. The solution to the reuse problem has built in ways of relieving the tension force before demolition. There are a few approaches we tried for efficiency: recycling, safe demolition, and self-repair of concrete.

A research project on demolishing pre-stressed tendons in concrete was funded by NSF. It is a very specific type of recycling. Demolition of pre-stressed concrete is difficult due to energy stored in the pre-stressed metal tendons which can be up to 125,000 pounds. Pre-stressed tendons originally bonded to concrete or grout may become debonded over time. Most of all flat plate building construction and bridge girders produced utilize these, so the problem is enormous and largely unresolved. The objective of this research was to develop and test means of re-bonding the tendons under stress for continued load transfer during use and most importantly during demolition. A breakable ring in a protective sleeve containing adhesive or shear struts designed to release adhesives were used. Both were designed to re-bond the tendons to the concrete with adhesive when the glass ring or strut breaks during debonding. Pre-stressed concrete bridge plates were made with adhesive embedded for repair at a concrete company in Illinois. Also, the self-repairing approach was applied to pre-stressed members by re-bonding the tendons to the concrete, should any become debonded. Adhesive-filled tubes placed on the tendons break when there is movement of the tendon against the concrete (see figure). The adhesive restores the bond in the full-scale deck plates that were tested in the field.

The basic principle of pre-stressing consists of producing internal compressive stresses in a structure before the working loads are applied by the application of tension from steel tendons. The resulting state of stress is much less than without the pre-stressing because the tension puts concrete into a higher state of compression so that further tension, as experienced in bending, will be negated. Even with long spans there should be no cracking. Pre-stressed concrete is lower in first cost and has less life cycle costs due to lessened need for maintenance, and therefore is now often preferred for cost and safety reasons. Despite its popularity, little research has gone into demolition problems that are becoming prevalent.

Few methods have been proposed for the safe and economical demolition and recycling of pre-stressed structures. The pre-stressed tendons act like rubber bands, with energy stored in the stretching process. During demolition of this type of concrete, the release of the stored energy may cause the tendon to be catastrophically released, threatening the safety of workers and the rest of the structure. This problem is magnified when the demolition is required in a dense urban setting. The proximity of the surrounding people and buildings must be considered when planning the demolition of pre-stressed structures. In addition, pre-stressed tendons originally bonded to concrete may become debonded over time. Most of all flat-plate building construction and bridge girders produced utilize tensioning of metal tendons, so the problem is indeed large.

This project concentrated on the pre-tensioning method of pre-stressing. In this method, steel wire strands, the tendons, are stressed in tension. While the stress is maintained in the tendons, concrete is poured around them. After the concrete has cured, the tension is released, and the concrete, which has formed a continuous bond with the tendons, is pre-stressed as the tendons attempt to shorten to the length they were.

After the stress is released, they are reloaded in the field. In service over time, the reinforcing may experience a "slipping" or debonding force between the concrete and the causing a corresponding change in the size of the cross-section of the tendon. There is some indication that cracking can occur at the bond of the tendon and any water intrusion could cause corrosion. The timely release of adhesives could counteract the loss of bond strength between the concrete and the pre-stressed tendon, thus increasing the stress transfer capabilities of the pre-stressed system and allow for safe demolition.

Consequently, the internal adhesive delivery system of re-bonding and repair chemicals is applicable to pre-stressed concrete. Construction of pre-stressed members was carried out with stiff hollow fibers containing adhesives. Over time, the relaxation of steel or concrete will cause shear that will fracture the glass tubes, release the adhesives, and cause bonding between the steel and concrete. This will help steel release the stress to the concrete at the time of demolition over shorter length of tendons and therefore with less energy.

Further the bonding between concrete and pre-stressing steel and concrete as a result of the internally released adhesives will also protect against corrosion of pre-stressed steel. This is very important because if a tenth of a millimeter is corroded away by rust, a loss of up to 13% of cross-sectional area in a 3mm diameter wire is incurred (Libby, 1984). Re-bonding over the life of the structure will also arrest the further deterioration of the structure over time. Such deterioration in the pre-stressed members includes cracking due to flexure, temperature, shrinkage, volume changes, and cracks at weak points as flange-web cracks.

The Illinois manufacturer of precast, pre-stressed bridge deck slabs donated some into which repair chemicals were placed. The company also assisted with the testing in the field. Due to their thin cross-section, these slabs with only one row of pre-stressed tendons were ideal in the testing of benefits provided by adhesive capsules located near the pre-stressed tendons within the slab.

This project was designed to test the ability of an adhesive to rebond pre-stressed steel tendons to concrete. The objective was to design a delivery system of adhesive to cracked concrete and debonded pre-stressed steel tendons under critical loads. The delivery systems being investigated were straight with U-shaped glass tubes filled with adhesive and fastened to pre-stressed tendons. Under stresses that may cause concrete cracking and debonding of the tendons from the concrete, the glass capsule would break and adhesive would be released in the critical area with the purpose of sealing cracks and rebonding tendons with the concrete.

The objectives of this research were to develop and test means of restraining or rebonding said tendons under stress for continued effective load transfer during use and most importantly for safe demolition. The investigated design is breakable shear struts designed to release adhesives for the bonded tendon application.

In pretensioned members, it is necessary to make sure the concrete or joint is bonded to the tendon so that if it is cut, there is no sudden release of energy. For newly constructed members, this would be done by placing brittle fibers containing an adhesive at certain points along the length of the tendon. These would be opened or released before demolition by shear or expansion during movement, releasing the adhesive that would form a bond. The member could then be cut at various places safely, releasing smaller amounts of the tendon's pre-stressed force. It is designed to yield the shortest debonding lengths to control the release of force. It is the difference between cutting one long rubber bands and cutting several short rubber bands.

A commercial polyurethane was used as the adhesive to fill multiple 5 mL straight test tubes. Glass tubing was also bent to produce u-shaped tubes which would be filled and fit around 3/8" steel tendons. Each u-shaped tube contained about 3.3 mL of adhesive. Purple dye was added to the adhesive to increase its visibility when tested, and fluorescent yellow dye was added to adhesive in later tubes so that it could be seen using a black light. Straight tubes and U-tubes were all sealed at the ends.

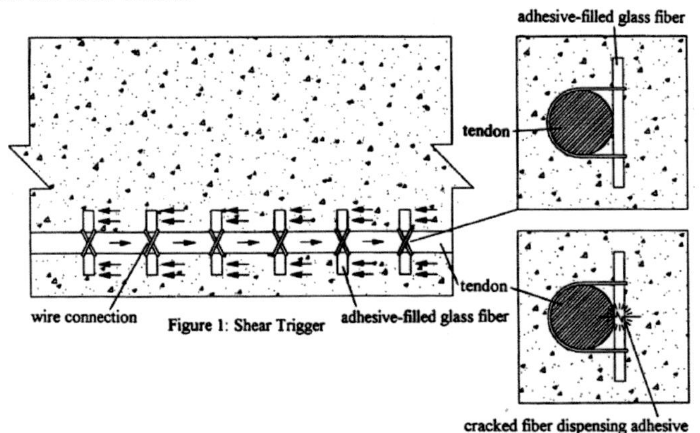

Drawing of repair adhesive tubes placed on the prestressed tendon.

CONCLUSIONS

The adhesive delivery system was tested in pre-stressed concrete slabs fabricated at a company. Full-scale testing in the field of simple bridge plates let us assess the viability of the system. We know that the adhesive survives as a liquid inside the tubes through the concrete placement and steam curing process and is released upon bending to failure. Sample cutting showed the adhesive to be liquid. The members with adhesive performed better in terms of strength retention and restoration of deflection. In addition, the cutting of the members showed that the samples without adhesive released the tendons with much greater force.

Photo of prestressed bridge plate specimens with self-repair tubes being poured in the field.

The experimental adhesive slabs performed the best at preventing retraction of the tendons into the slab, which indicates that the tendons were rebonded. Because the direct bond between the concrete and the tendons prevented much retraction of tendons, the mechanical loadings and the cutting of the slabs did little to debond the tendons from the concrete. However, at a larger scale in members that are very long, the retraction might be more pronounced. Even the observed slight debonding of tendons could lead to corrosion of the steel and the addition of the self-repair system would aid in lessening the retraction until the member could be fully repaired.

The experimental slabs with the repair tubes were strongest after many strength tests thus indicating the extra advantage of self-repair.

CHAPTER 8:

CO_2 ABSORBING COATINGS

All of my solutions for preserving concrete address porosity and damage prevention. Concrete preservation again is important because of the widespread use of it and the incredible impact its manufacturing has on the environment. If we can use less (because we make it more durable) we can begin to mitigate the problem.

If we strengthened concrete and closed off the pores to limit penetration by chemicals and water while at the same time taking up carbon dioxide (which reacts with the basic cement component – lime) to make the basic glue in cement CSH, we could limit the outside energy required and extend durability. This will work with extant concrete to revive its strength and increase its lifetime all the while taking up CO2. Also, releasing oxygen in the process is desirable for environmental reasons. I have developed a coating that does all of this.

Materials used in buildings and roads that can improve the environment are the solution to too much Co2 and deteriorating concrete infrastructure. I aim to produce a concrete that undoes environmental strain and yields a net environmental asset by a) sequestering CO2 in concrete which is densified and made more durable by forming a coating that prevents deterioration, b) giving off oxygen, and c) catalytically converting NOx, SO2, CO2 to acids which are then neutralized by the very basic concrete.

SELF-STRENGTHENING CONCRETE COATING

The coating material consists of an inexpensive silicate that reacts with the cement then continues to densify and strengthen over time by reacting with CO_2 from exhaust and atmosphere (in the presence of water vapor). Over time, these concrete structures will deteriorate. This technology focuses on self-repairing this concrete via CO_2 absorption, extending useful life while preventing damage such as corrosion.

At the tailpipes of gas-burning vehicles, there is a uniquely high concentration of carbon dioxide (CO_2) and water vapor (H_2O). Each gallon of fuel burned yields 19.6 pounds of CO_2. Our conducted tests, which used only the amount of CO_2 equivalent to that available in the atmosphere, showed great promise for the development of this enhanced cementitious material for roads designed to take up that CO_2 quickly to strengthen the underlying concrete structure, and make it less permeable to water and chemicals. Our field research focused on creating a novel cementitious repair material in bridges that reacts with the CO_2 and H_2O from the actual atmosphere to form a strong protective top layer within the concrete pores, and allow the coating material to migrate, react, and protect against damage by filling in any pores.

We used a silicate source to make CSH (the "glue") in concrete for internal strengthening because it will produce X(OH) in-situ, which will migrate throughout the concrete, react with CO_2, and then afford protection against chemical intrusion. My tests showed that this cementitious material with a silicate coating exposed to CO_2 will prevent compression cracking and the intrusions of environmental elements into concrete by increasing strength and density, reducing porosity, and raising surface hardness. What was shown in the first set of samples in chambers containing 5% CO_2 in the presence of water vapor was that exposure of this material to the equivalent of atmospheric CO_2 over two years makes it 2.4 times stronger.

Depth of carbonation was 5/8 inch, density was increased 10 % and porosity reduced 6% percent. These values went much higher with better lab tests and at field tests in the actual atmosphere. Field samples exposed only to the atmosphere, with no special CO_2 concentration, absorbed 0.0035 moles of CO_2 per 1 g. of coating material and became 3.9 times stronger in compression in the 4 months of atmospheric exposure. The coating penetrated the concrete approximately 1 mm. in a year under normal atmosphere.

The tests in the field reacted totally, 100% with atmospheric CO_2 in a four-month period. This was much more rapid than expected. We postulate that the higher water vapor content and perhaps temperatures pushed the reaction to faster completion. The coating was designed to carbonate all the way through the coating or patch. When the carbonation reaches completion, another coating would restore the ability to absorb CO_2, and strengthen, densify, increase hardness, reduce porosity and prevent other internal damage such as corrosion.

This investigation was into the development of this innovative retro-fit coating material to extend the service life of airport runway pavements, by taking up CO_2 to make in-situ protection while strengthening the concrete itself. There are several silicates which could be used.

The coating, which is to be placed on existing concrete surfaces, consists of a solution

(XSi_5O_{11}), which reacts with the free calcium in the extant concrete to produce calcium silicate hydrates (C-S-H) and hydroxide. This resultant hydroxide then reacts with the CO_2 in the atmosphere to create an effective compound, a carbonate in silica gel network . C-S-H is the primary bonding compound, or "mineral glue," in the cement. This resultant C-S-H reacts with the CO_2 in the atmosphere to yield a dimensionally stable silica gel network with strong calcium carbonate ($CaCO_3$) in its interstitial pores. We have proved that this chemistry works. This stronger, denser "crust" reduces porosity, strengthens and densifies.

Experiments on the chemistry of the reaction have verified that 1) the XSi_5O_{11} reacted with the lime in cement to produce C-S-H. This means that the XSi_5O_{11} would react with the lime in any concrete 2) this C-S-H reacts with the CO_2 gas in the presence of water vapor forming $CaCO_3$. The formation of C-S-H creates volume and reduces porosity and the silica left behind when the C-S-H reacts with the CO_2 further fills in the microstructure.

The following general reactions demonstrate the formation of C-S-H. Lime, calcium hydroxide (Ca $(OH)_2$), is the source of calcium (Ca^{+2}) in these reactions.

Reaction 1: H_2O

$XSiO_3$ + $xCa(OH)_2$ → $xCaO.ySiO_3.H_2O$ + $2yXOH$)

X silicate *calcium hydroxide* → *calcium silicate hydrate* *X hydroxide*

Reaction 2: H_2O

$xCaO.ySiO_3.H_2O$ + $2yX(OH)$ + xCO_2 → $xCaCO_3$ + $ySiO_3$ + XCO_3

$ySiO_3$ + H_2O

calcium silicate hydrate *hydroxide carbon dioxide* → *calcium carbonate* *silica gel network*

The silica network maintains the same overall volume as the original C-S-H, while the material gains density and strength due to the $CaCO_3$ precipitation in the network's voids. It creates a denser, more volumetrically stable final material in a very short time frame. It was predicted that the initially elevated pH resulting from the application of the basic silicate coating will increase the solubility of the silicates, thereby driving reaction 1 and insuring that Ca $(OH)_2$ reacts with the silicate (reaction 1). That was proven, see figure. No evidence of calcium is seen either in the form of the starting material Ca $(OH)_2$ or $CaCO_3$. From this fact, we conclude that all of the calcium in $Ca(OH)_2$ has reacted with the silicate solution to form the C-S-H phase.

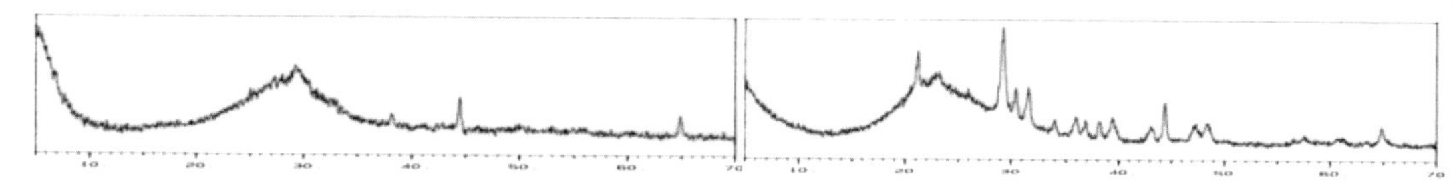

Left, XRD pattern of C-S-H formed by XSi501 and $Ca(OH)_2$. *Right, XRD pattern of CaCO3 formed after C-S-H was exposed to CO2.*

LAB TESTS

The results of this run of experiments yielded a variety of evidence that the C-S-H/sand mixture did absorb CO_2. The control samples exposed to nitrogen dissolved in water and the CO_2 cured ones, even after only 10 minutes in the tent with CO_2, remained intact in water. The CO_2 exposed samples became very hot (about 110° F) just after they were exposed to CO_2 with the controls not getting hot. This heat reaction indicated a chemical reaction. Furthermore, the CO_2 exposed samples had a harder surface. The samples exposed to 100% CO_2 for 24 hours gained an average of 10% additional dry weight while the control samples exposed to nitrogen gained virtually no weight (less than 0.8%).

The fifth type of evidence was the difference in compressive strength. The compression tests consisted of using a compressive strength-testing machine. The samples were compressed until failure. It was seen that the samples exposed to CO_2 had an average of 136% higher strength capacity i.e., they were approximately 2.4 times stronger, than the control ones. Additionally, porosity tests were carried out using kerosene was used in the experiments. It was observed that there was an average of 6% decrease in porosity of the samples that were exposed to 100% carbon dioxide for 24 hours.

EXPERIMENTS AND RESULTS

Tests performed on the first set of samples revealed that there was a direct relationship between CO_2 exposure and compressive strength, i.e., as the samples absorbed CO_2, they increased in compressive strength. The four-month field samples were compressed until failure. The direct relationship between CO_2 absorption (carbonation) and compressive strength is again seen. Samples that were exposed to 24 hours of 100% CO_2 had 70% carbonation, and gained an average of 136% compressive strength. They were approximately 2.4 times stronger in compression than the control ones. Field samples that were exposed to

4 months of atmospheric CO_2 had 100% carbonation and gained an average of 288% compressive strength i.e., they were approximately 3.9 times stronger in compression than the control ones. Field samples that were 6 months old showed 100% carbonation.

Carbonation depth tests were performed on compacted CSH samples. The samples were cleaved first, and then phenolphthalein was sprayed on. The phenolphthalein stain turned white and indicated that the CO_2 had penetrated to 5/8 inch in the samples exposed to 100% CO_2 for 24 hours. The nitrogen exposed samples showed no carbonation.

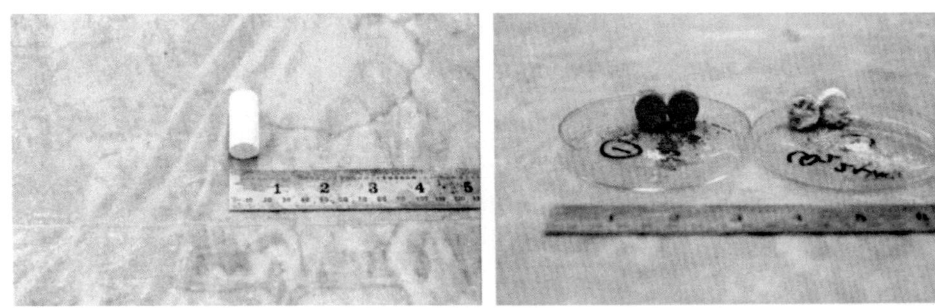

Left, Photo of compacted C-S-H sample Right, Samples coated with phenolphthalein, left is CO2 exposed CSH sample that turned bright pink indicating carbonation, and on the right is the sample exposed to nitrogen, showing no carbonation.

In-Field Tests

The amount of CO_2 absorbed by the samples was calculated and showed absorption.

Group	Exposure	CO_2 absorption Moles/1 g. sample	Compressive strength N/mm²	Psi	%Change	Porosity %
1	Nitrogen (control)	0	0.5661	82	0	44
2	24hours100%CO_2	0.00236	1.3380	194	136	38

This experimentation produced these findings: the material that forms, after C-S-H reacts with atmospheric CO_2 for the equivalent of two years, is 2.4 times stronger, is denser and less porous and forms a hard 5/8 inch thick crust. These experimental findings support the hypothesis that the formation of C-S-H reduces porosity and when the C-S-H reacts with the CO_2, it fills in the microstructure.

COATING ON FULL SCALE BRIDGE DECKS

Full scale bridges made for a prior Transportation Research Board IDEA project were used in these coating experiments. XSi_5O_{11} as a coating, and C-S-H as a patch material were applied on the bridge decks in 5 feet long and 6 inches wide strips.

Photo of the color changes that have taken place after 6 months of application. Surface, where XSi5O11 was applied as a coating, got a transparent color, on the left, and surface, where C-S-H was applied got whiter, in the middle. Bridges in winter on the right.

XSi_5O_{11} was applied in two ways. It was either applied directly, or after mixed with cab-o-sil. Cab-o-sil can suspend in solutions, therefore, it was used to create a thicker and denser coating. C-S-H was applied either directly on the deck surfaces, or over the XSi_5O_{11} coatings. C-S-H was mixed with sand before the applications to reduce the shrinkage cracks that it may undergo. After three months, observations were made. It was seen that there was a color change where the materials were applied. The strips of the decks where XSi_5O_{11} was applied had taken a transparent color. The ones where C-S-H was applied have become whiter. The color change means the reactions have taken place.

APPLICATIONS ON BRIDGE DECKS

Four full-scale bridges, constructed out of 20 N/mm^2 (3000 psi.) strength concrete were used. Two strips on both sides of each deck that were 1.8 m. (6") wide were painted with the applied materials. The materials were applied as a coating. Every foot on the strip indicated a new three months application. The figure shows photos of the color changes that have taken place on the deck surfaces after six months.

MICROSCOPY ANALYSES

Core drilling was performed on the bridge decks. 50 mm. (2") diameter core samples were taken from the following surfaces: (a) surface coated with XSi_5O_{11}, (b) surface with C-S-H and sand mixture, (c) surface coated with XSi_5O_{11} and then with C-S-H and sand mixture, and (d) surface that had no application. It is seen that there were fewer cracks and pores in all of the first three surfaces compared to the control one. Also, it was observed that the most reduction of cracks and pores and increase in density occurred in all samples. Furthermore, a transition zone, which was approximately 1 mm. deep, was detected on the surface of the first specimen. This is taken as a zone of densification.

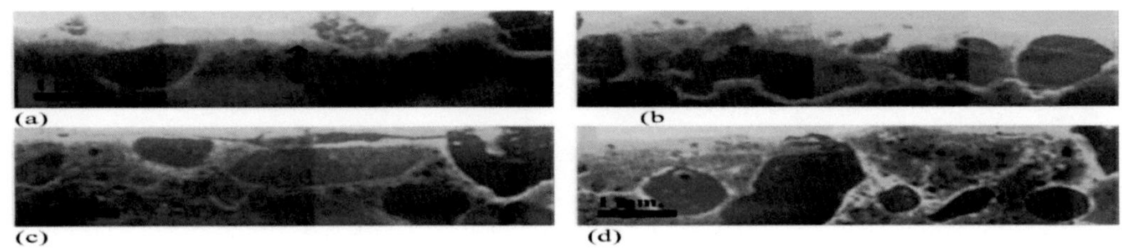

Microscope photos of the deck surfaces: (a) surface coated with XSi5O11,(b) surface patched with C-S-H and sand mixture, (c) surface coated with XSi5O11and then with C-S-H and sand mixture, and (d) surface that had no application.

TESTING OF COATINGS

Concrete is a very common construction material found in buildings, roads, bridges, and sidewalks. The objective of this work is to test the effects of two different coatings designed specifically for concrete. These coatings are designed to uptake CO_2 outgassed by the concrete and in return produce O_2 which benefits the environment.

The two coatings have a common chemical: lithium polysilicate (LPS). When LPS is mixed with a variety of chemicals it is known to uptake CO_2 while releasing O_2. There has been some speculation on whether CH_4 (methane) gas is also released in these reactions. In all studies performed the level of methane produced has been on a scale of two orders of magnitude less than the initial CO_2 level and the level of O_2 produced. The first coating is 1:1 mixture by weight of LPS and lime (CaO/CaOH). When LPS and lime are mixed together they react to make CSH which then can react with atmospheric CO_2 to produce calcium carbonate ($CaCO_3$) a hard, rock-like structure, water (H_2O) and oxygen (O_2). The coating is spread over the top of the concrete block. Initially, the coating is thick but quickly hardens.

The second coating is a once again a 1:1 mixture by weight of LPS and this time with calcium peroxide (CaO_2). Calcium peroxide does not react with LPS by itself, but when it meets atmospheric water it degrades into lime which then reacts with LPS in the same manner as explained above. The initial coating is a thick, viscous mixture but hardens into a similar coating as the LPS and lime. The readings taken for each trial were graphed and the results are shown in the figures.

As shown by the graphs, the amount of CO_2 decreases exponentially as a function of time and O_2 increases. Both also show a leveling off effect. For CO_2, it approaches zero but has

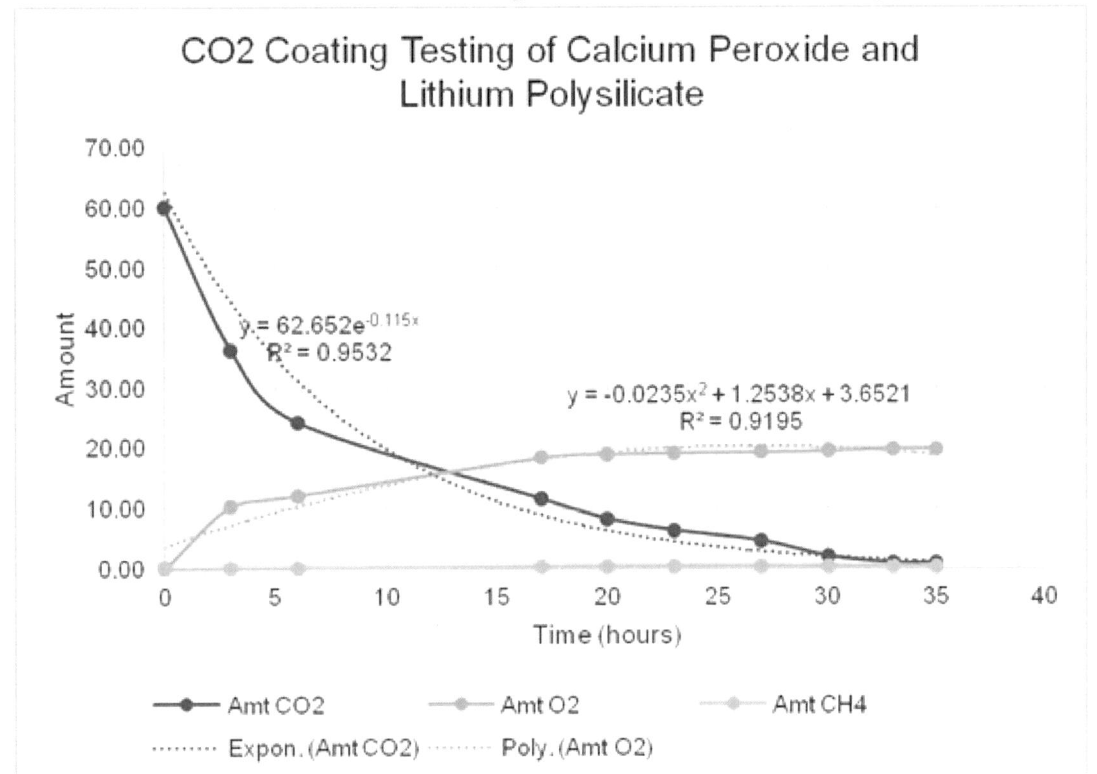

never reached that value and O_2 levels off as it approaches 20. The amount of CH_4 in both graphs evens out at approximately 0.25. The leveling off effect probably means that the amount of oxygen produced is dependent on the amount of available carbon dioxide. As the reaction proceeds less carbon dioxide is present, therefore less oxygen is produced.

What is unexpected between the two sets of data is that the LPS and calcium peroxide coating initially took up CO_2 much faster than the LPS and lime coating. This is shown by the slope value of each equation with the slope of the LPS and calcium peroxide being significantly larger than that of the LPS and lime (62.652 compared to 46.95). The reason this is unexpected is because calcium peroxide must first be converted to lime to react with the CO_2. However, with this being the case it raises questions about whether the degradation of calcium peroxide by water also takes up CO_2.

The effect of the coating (if any), on the structural integrity of the concrete needed study. The lime and LSH coating applied reacts with the concrete and makes it stronger.

INCREASED SURFACE AREA

The idea was to put holes in the cement samples so that the increased surface area would absorb more CO2. A comparison was made to samples without holes and with no coatings. Six sample types were made and place in a tank filled with CO2 then a compression test was run on them. Finally, a phenolphthalein test to ascertain the uptake of co2.

The samples without holes were weaker and exhibited less CO2 uptake. For samples with holes, the maximum peak load for sample 6 was 4061.4 lb. over 0.25 in. This peak load value was greater than the peak load value of the comparison sample without coating by about 1000 lb. This indicates that the addition of all three additives coating in this amount increased the strength of the sample.

After phenolphthalein was dispersed onto the sample there was a color change. Both the outside edges and the middle of the holes did not show a color change, indicating a reaction occurred of absorbing CO_2. The purple that was visible was very light in color and was a little spotty.

After evaluating all of the samples, a common pattern was observed with respect to the phenolphthalein test. Both the outside edges and the middle of the holes had no color change, indicating that those spots underwent the reaction of converting CO_2. This suggests that adding holes to the samples increased the area in which the samples underwent the reaction.

Sample	**additive(s)**	**% CO_2 decrease**	**% O_2 increase**	**peak load (lb)**	**% strength increase**
1	None (no coating)	11.1	59.5	3071.2	N/A
2	CaO_2	12.8	418	4905.1	59.7
3	CaO_2, LPS, lime	11.7	143	4075.3	32.7

Sample	additive(s)	% CO_2 decrease	% O_2 increase	peak load (lb)	% strength increase
4	CaO_2, LPS, lime	9.84	126	3667.3	19.4
5	CaO_2	12.07	210	4800.1	56.3
6	CaO_2, LPS, lime	13.37	156	4061.4	32.2

In addition, the strength of the samples did increase with the addition of the holes. The samples containing only calcium peroxide had the greatest percent increase in strength. This is not unexpected as calcium peroxide absorbs both CO_2 and moisture from the air. As this happens, a reaction takes place that fills the pores in the cement with lime deposits, making the sample denser and therefore stronger.

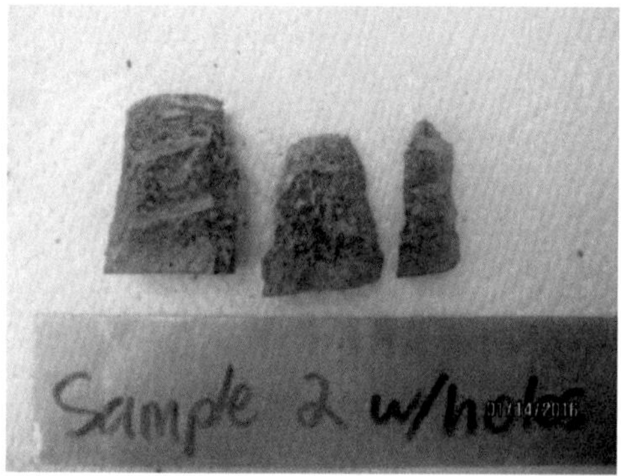

When comparing the CO_2 test among all of the samples, sample 2 increased the O_2 levels by the greatest percentage (510%). All of the other samples decreased CO_2 levels by roughly the same percent.

CO_2 ABSORPTION WITH ACCELERATED TIME

In previous testing it has been confirmed that concrete can be used to absorb CO_2 when certain surface coatings are applied. These coatings consist of a combination of lithium polysilicate ($Li_2Si_5O_{11}$) calcium peroxide (CaO_2) a lime precursor or Lime and water. Also, just calcium peroxide and water are used. The calcium peroxide and water are mixed in a 1:1 weight ratio. When the coating containing lime (or lime produce by calcium peroxide) and lithium polysilicate is exposed to water vapor it reacts in a way that produces a compound called CSH. CSH is the 'glue' of cement but also acts as a CO_2 absorbent, taking up CO_2 from the atmosphere while producing oxygen.

The coating with just calcium peroxide and water reacts with CO2 and gives off oxygen while converting into lime. The coating made with lithium polysilicate and calcium peroxide there is an additional step added to the reaction mechanism. The calcium peroxide, when exposed to CO_2 in proper humidity levels of nearly 50%, reacts with the CO_2 to form lime and release oxygen. That lime is also exposed to the lithium polysilicate and the reaction of lime and lithium polysilicate in humidity ensues in which CSH is produced, and then the further reaction of CO_2 absorbed, oxygen produced and calcium carbonate made, which is the hard material in concrete.

The goal of this testing was to test how much of the CO_2 absorbed is the result of calcium peroxide converting into lime and how much was resulting from the production of CHS by the lime reaction with silicate and moisture. The tests were modeled on previous work in which samples were prepared with various coatings and placed in an air chamber at nearly 100% CO_2 by volume. CO_2, O_2, and humidity levels were then monitored. Since CO_2 absorbance is a function of the CO_2 Vol. % in the chamber, efforts were made to keep the chamber at high CO_2 levels. This was done every 24 hours by 'purging' the air chamber, meaning refilling it to near 100% CO_2. These tests were run for 3 days.

The first sample had a coating of calcium peroxide and water. This was used to measure the effects of only the reaction where calcium peroxide uses CO_2 to convert into lime. Next a sample with a coating of lithium polysilicate, calcium peroxide, and water was made.

These numbers were then compared to determine the effectiveness of both reactions separately in terms of CO_2 absorbed and oxygen released. Later the results were compared to two controls, one with an uncoated cement sample in the chamber, and another with the chamber empty. The concrete used in this testing was Portland cement, and 2 by 2 by 2 inch cubes were prepared for coating. This means 24 square inches were coated.

TEST RESULTS

Sample	Mass CaO2 (g)	Mass Li2Si5O11 (g)	Mass H2O (g)
1	100	0	100
2	100	20	100

This figure shows the CO_2 being absorbed, as compared to O_2 production. The sudden spikes in the data show when the air chamber was purged and refilled with 100% CO_2. This was at 24 and 48 hours. The figure shows that in times between the purges CO_2 levels fall and the Oxygen levels rise. This relationship is evidence of the reaction functioning properly.

Sample 2 had lithium polysilicate, calcium peroxide and water and the additional step in the reaction (peroxide converting to lime) resulted in a larger consumption of CO_2 and oxygen release. This is evident in the first day of the testing when sample 1 converted 8.8 % vol. of CO_2 in the first day when sample 2 converted 14.6 % vol. of CO_2. These numbers convert to 23.672 L and 39.274 L respectively. In the first day of testing, the coating from sample 2 absorbed 15.602 L of CO_2 more than sample one. Similar calculations can be done to show that the coating of Sample 2 was more effective over the 3 days. This control sample chart shows CO2 absorbed and O2 produced. The CO2 levels go down between 24 hour purges while the O2 levels go up in between purges.

	Mass CaO 2 (g)	**Mass Li2 Si5O11 (g)**	**Mass H2O (g)**	**CO2 absrb (L)**	**CO2 absrb (mol)**	**O2 (L)**	**O2 (mol)**	**Peak stress (psi)**	**Peak Ld (lb)**
1	100	0	100	44.654	1.93	3.04	.131	757	3028
2	100	20	100	66.712	2.85	1.45	..063	597	2388
3 control	0	0	0	30.397	1.31	0.565	.024	1013	4052

Summary table of totals of CO2 absorbed and O2 released over a 3-day period for coated and one uncoated sample. It also shows the strength after 3 days in high co2 exposure

CONCLUSION

Both reactions proved capable of absorbing CO_2. Within the first day in about 90% CO2 by volume, it is estimated that the first reaction (just calcium peroxide and moisture) will absorb 23.672 L of CO_2 per every 24 square inches of coating. Over the entire 3-day period the coating on sample 1 converted a total of 44.654L.

Based on this information 1.93 moles of CO_2 were converted over the course of the three days. Comparing the fact that one day in the CO2 chamber is equal to 2 years we can estimate that 3 days is 6 years and that is the volume of co2 gas absorbed by that coating. The second reaction (lithium silicate and calcium peroxide to make CSH which absorbs $CO_{2)}$ is estimated to take in 39.274 L of CO_2 per every 24 square inches of coating. which is 2.85 moles.

Sample 1 with calcium peroxide and water out performed sample 2 in terms of oxygen produced in the reactions and in strength after 3 days in a CO2 filled chamber. However, sample 2 with calcium peroxide water and lithium silicate absorbed 1/3 more CO2 over 3 days.

The control with no coating absorbed much less CO2 and released much less oxygen than the samples with the coatings However it was stronger I questioned this so I ran more experiments and it was weaker.

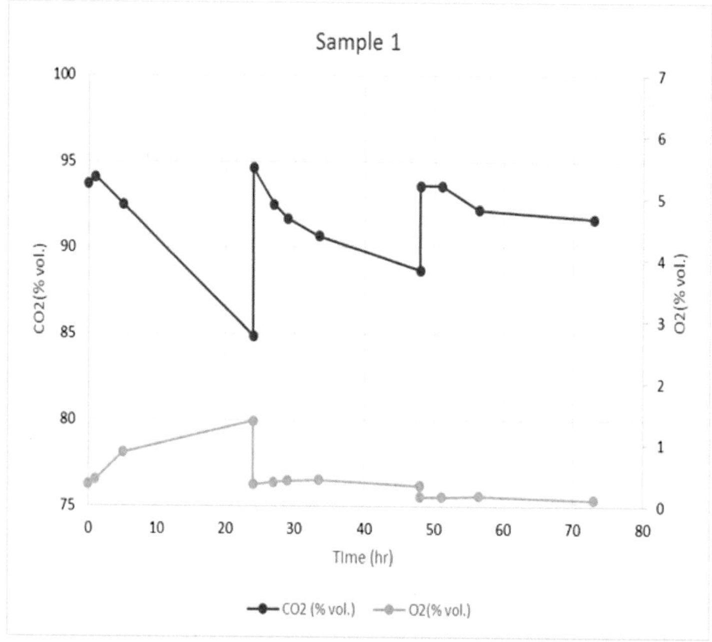

Sample 1 reactions over time shows the CO2 being absorbed, as compared to O2 production. the CO2 levels go down 10, 6 and 3% between 24-hour purges while O2 levels rise in between purges by 1.03,.08, and .02%

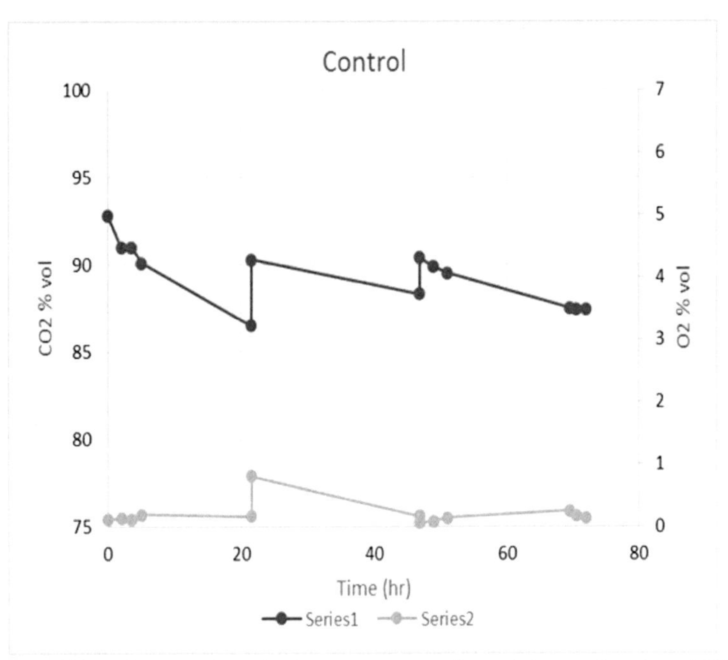

This figure shows the CO_2 being absorbed, as compared to O_2 production. The sudden spikes in the data show when the air chamber was purged and refilled with 100% CO_2. This was at 24 and 48 hours. The figure shows that in times between the purges CO_2 levels fall and the Oxygen levels rise. This relationship is evidence of the reaction functioning properly.

Sample	Mass CaO2 (g)	Mass Li2 Si5O 11 (g)	Mass H2O (g)	CO2 absrb (L)	CO2 absrb (mol)	O2 release (L)	O2 release (mol)	Peak stress (psi)	Peak Ld (lb)
1	100	0	100	44.654	1.93	3.04	.131	757	3028
2	100	20	100	66.712	2.85	1.45	..063	597	2388
3control	0	0	0	30.397	1.31	0.565	.024	1013	4052

Summary table of totals of CO_2 absorbed and O_2 released over a 3-day period for coated and one uncoated sample. It also shows the strength after 3 days in high co2 exposure

CATALYTIC CONVERSION WITHIN CEMENT

The innovation here is a retro-fit coating material for concrete infrastructure applications that repairs itself by taking up greenhouse gases and catalyzes them. These gases include carbon dioxide, water vapor, nitrous oxides, and sulphur dioxide.

Most paints are made from or latex and cover miles of surfaces, emitting toxins as they dry. This is an opportunity to absorb gases or act as a catalyst for potentially profitable reactions. A catalyst, in popular, terms means change agent. A more scientific definition is one that is non-diminishing and endlessly-acting. These paints may be for concrete using the same chemistry as above with additional chemicals (such as titanium dioxide, a catalyst when exposed to sun).

Specifically, this coating is sustainable because it will lengthen the life of concrete, reduce the need for new concrete structures, and will reduce the release of CO_2 dramatically due to the coating. Its main draw is the ability to catalyze and transform CO_2 and other greenhouse gases in vast volumes into harmless liquids (acids which are then neutralized by the basic concrete).

I used titanium oxide doped with nitrogen, sulfur or carbon in the silicate matrix which transform CO_2, water, NOx, and SOx into acids which are then neutralized by the calcium carbonate present in the densified concrete in and under the coating.

I used prior lab research on making coatings from silicates (which are like zeolites with a lattice formation) which is designed to self-strengthen and densify by taking up CO_2 from the air in the presence of water vapor. I learned from Asahi and Anpo 1o add titanium oxide particles that are doped with nitrogen to react better in ambient light to photocatalyse NOx. I photocatalysed three greenhouse gases – CO_2, and H_2O, NOx and SOx – by doping the titanium oxides with all of N, S, and C. From the paper on coating with calcium carbonate particles, I learned that when the reactions produce acids they can be neutralized by the calcium carbonate. In contrast, our coating will already have the calcium carbonate available in the concrete and coating after CO_2 densification to neutralize the acids.

In sum, the innovation is a concrete coating made of inexpensive silicate and lime that makes CSH which then reacts with CO_2 to fill in the pores of the concrete while not filing in the pores of the coating. It is a special type of a mineral coating. These "bind within the substrate, forming a microcrystlaine structure and become intergral to the coated substrate" but this one forms an interstitial lattice work. The titanium nano particles are in the coating but do not react at all with the underlying concrete. They photocatalyze the gases that come into the intersitial areas in the coating. The reacted gases become acids in the presence of water and are netralized by the calcium carbonate in the coating or concrete.

INVESTIGATIONS WITH TITANIUM OXIDE

The purpose of these experiments was to determine if a coating of titanium (IV) oxide and polysiloxane on the outside of a concrete specimen as well as a coating of lithium polysilicate and lime on the inside of a concrete specimen would absorb greenhouse gases and convert the trapped gases into non-harmful byproducts (again acids) which are then neutralized by the very basic concrete.

The materials used in these experiments can be separated into two parts. The first group was the materials for manufacturing of the concrete specimens. The second group was the materials used in the coatings. The concrete consisted of distilled water, cement, and rocks (non-alkaline). The coating contained titanium (IV) oxide, polysiloxane paint, lithium polysilicate, and lime.

A standard concrete mold was used. In order to get the desired 2 in x 2 in x 2 in specimen geometries, cardboard spacers were cut and placed into the mold. Using the standard mold, there were fifty 2 in. x 2 in. specimens. The mold with spacers was then sprayed with cooking spray, which acts as a release for the concrete.

The concrete specimens were made with a 1:1 ratio of concrete mix to distilled water by volume. To properly mix the concrete and distilled water, a blender was used until the resulting mixture was well blended. After the concrete was mixed, the alkaline-free rocks were added until the rocks were coated with concrete (typically about 2:1 rocks to concrete mix, by volume). The concrete was placed into the mold in the desired 2 in x 2 in x 2 in dimensions. The specimens were then left for approximately 24-48 hours, until the specimens were cured enough to remove from the molds. They were placed in a water bath for 28 days. After the water bath, placement on a screen dried the specimens. The screen was used so all four sides of the specimen were dried. After the specimens were dry, approximately 24 hours later, they were then ready for coating.

The inner coating was intended to make the calcium carbonate form taking up CO2. It consisted of lithium polysilicate and lime mixed at a 1:1 ratio, by weight. The mixed coating was then put into a needle and syringe. The needle and syringe were used to inject the concrete specimens in three central locations. The coating was injected into each specimen as far into the porous specimen as possible. After the inner coat was applied and dried, the outer coat was applied to the specimens.

The outer coating of titanium oxide and polysiloxane was intended to take up CO2 and make carbonic acid. It was mixed at a 1:1 ratio by volume. This coating was applied using a paintbrush. All sides were coated, and this process was repeated to ensure a good, even coat.

The specimens were tested after they were prepared and coated. The testing was split into two groups. The first group (control) was placed under nitrogen to avoid any co2 uptake. The second group was placed outside in a CO_2 rich environment (a bus parking lot). The second group served as the experimental specimens. Within each group, the specimens were made as controls (no coatings), outer coating, or both inner and outer coatings. The controls were specimens without any coating placed on them. The outer coating specimens only had the titanium oxide and polysiloxane coating. The specimens with both coatings had both the titanium oxide and polysiloxane on the outside with the lithium polysilicate and lime coating on the inside.

Group	**Type**	**Number of Specimens**
Nitrogen on CO2	Control	8
	Outer Coating	7
	Both Coatings	6
Outside with CO2	Control	9
	Outer Coating	9
	Both Coatings	9

After four months of conditioning outside or under nitrogen, the amount of SO_x and NO_x uptake was ascertained. Sulfate test strips (test strips that approximate SO_4 content from 0-500 ppm) and Nitrite/Nitrate test strips (test strips that approximate NO_2/NO_3 content from 0-20 ppm and 0-200 ppm, respectively) typically used for testing aquarium water or drinking water were used to evaluate the presence of sulfates, nitrites, and nitrates.

Each concrete specimen was ground using a mortar and pestle. Dry powder (5.117 g) was weighed into a cup, with approximately 8 g of rock reinforcement. Distilled water (40 mL) was then added to the mixture. The mixture was stirred three times over 30 minutes. Following this, the testing was conducted.

Conditioning	Coating	Nitrite (ppm)	Nitrate (ppm)	Sulfate (ppm)
Environmental	None	0.5	10	300
	Outer	0.5	10	300
	Outer/Inner	0.5	2	300
Nitrogen	None	0.5	5	500
	Outer	0.5	5	500
	Outer/Inner	0.25	2	500

Phenolphthalein Tests to assess amount of CO2 taken up were conducted using a hammer and chisel. Phenolphthalein dye was applied to cross-section through a syringe.

Specimen Coating(s)	Conditioning	Phenolphthalein Test Results
None	N_2	Fuchsia, entire, immediate
Outer	N_2	Fuchsia, entire, immediate
Outer/Inner	N_2	Fuchsia, entire, immediate
None	Outside	Clear, slight pink hue after 24 hours
Outer	Outside	Clear, slight pink hue after 24 hours
Outer/Inner	Outside	Clear, pink circle in center of cross-section (penny-sized)

The outside coating (titanium in a porous paint) was expected to be able to absorb CO_2 and especially SO_x and NO_x. The inside coating was designed for CO_2 absorption and should neutralize acids formed by the outside coating. Thus, the samples that were exposed outside that had both coatings should have been able to absorb both SO_x and NO_x as well as absorb CO_2 with the outside coating and CO_2 with the inside coating. These samples had the biggest change from alkalinity so it is logical that they probably absorbed the most gases. The samples in the air controlled air chambers without any gases other than nitrogen were the most alkaline or pink with phenolphthalein and therefore, the least changed and absorbed the least amount of gases.

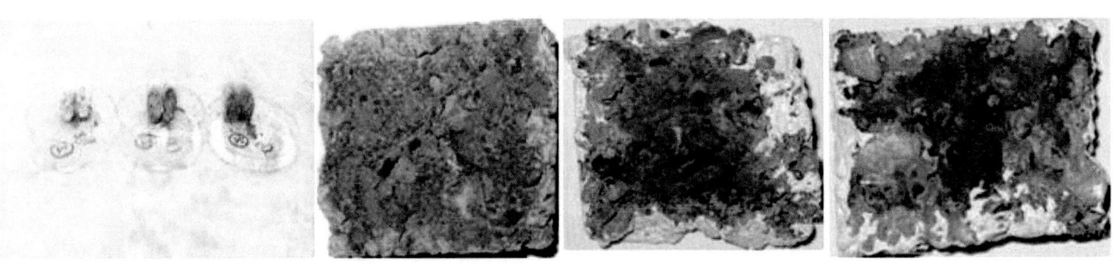

Left, samples coated with phenolphthalein; the first is an exposed sample, white, indicating carbonation with the next being samples exposed to nitrogen, pink, showing no carbonation. Next. three photos 15 days after fabrication in ambient air samples with a silicate calcium coating and additional mineral catalyst additive have deep carbonation. Phenolphthalein tests of left control specimen. Next is the coated specimen, and right is the specimen coated with catalyst additive.

Half of each concrete specimen was ground up using a mortar and pestle. Concrete dust was then added to distilled water (40 mL) and allowed to soak for 1 minute. Solution was decanted into a beaker and a pH test strip was submerged for a reading.

Specimen Coating(s)	**Conditioning**	**pH Test Estimate**
None	N_2	11.5
Outer	N_2	11.5
Outer/Inner	N_2	11.5
None	Outside	8.8
Outer	Outside	9.2
Outer/Inner	Outside	8.4

Sulfate Tests: Each concrete specimen was ground using a mortar and pestle. Dry powder (5.117 g) was weighed into a cup with approximately 8 g of rock reinforcement. Distilled water (40 mL) was then added to the solution. The solution was stirred three times over 30 minutes. Test strip submerged in the solution and the color measured against the Sulfate scale. The same procedure was undergone for Nitrite/Nitrate levels.

Conditioning	Coating	Nitrite (ppm)	Nitrate (ppm)	Sulfate (ppm)
Environmental	None	0.5	10	300
	Outer	0.5	10	300
	Outer/Inner	0.5	2	300
Nitrogen	None	0.5	5	500
	Outer	0.5	5	500
	Outer/Inner	0.25	2	500

Summary of Nitrite/Nitrate/Sulfate Test Results for 6-mo samples.

TEST CONCLUSIONS

Gases as CO_2, SO_x, and NO_x are absorbed from the atmosphere and become an acid that is subsequently neutralized by the concrete's inherent alkalinity. As a result, decreases in a concrete's alkalinity are associated with increases in gases absorbed from the atmosphere. The concrete with both an outer and inner coating demonstrated the lowest alkalinity, and consequently, an inner and outer coating maximizes pollutant neutralization. Sulfate testing suggested decreased levels of sulfate present in specimens conditioned outside.

Nitrate testing demonstrated higher nitrate levels in specimens conditioned outside while nitrite testing was too crude to accurately infer differences in nitrite levels between specimens.

The outside coating (titanium in a porous paint) was expected to be able to absorb CO_2 and especially SO_x and NO_x. The inside coating was designed for CO_2 absorption and should neutralize acids formed by the outside coating. Thus, the samples that were exposed outside that had both coatings should have been able to absorb both SO_x and NO_x as well as absorb CO_2 with the outside coating and CO_2 with the inside coating. These samples had the biggest change from alkalinity so it is logical that they probably absorbed the most gases. The samples in the air controlled air chambers without any gases other than nitrogen were the most alkaline or pink with phenolphthalein and therefore, the least changed and absorbed the least amount of gases.

Phenolphthalein tests were conducted of control samples, samples with only an outer coating, and samples with both an inner and outer coating. There appeared to be some indications on the outer perimeter of carbonation in samples with the inner/outer coatings. No indication of carbonation in the control samples. The depth of carbonation was assessed in all samples.

References cited

Asahi,R, and Anpo, M, Titanium compounds and their impact on air pollutants transformation
https://journals.agh.edu.pl/geol/article/downloadSuppFile/1304/754

Anpo M., 2000. Utilization of TiO_2 photocatalysts in green chemistry.
Pure and Applied Chemistry, 72, 1787–1792.

Asahi R., Morikawa T., Ohwaki T., Aoki K., Taga Y. 2001. Visible-light photocatalysis
in nitrogen-doped titanium oxides. *Science*, 293, 269-271.

CHAPTER 9:

SELF-FORMING POLYMER/CERAMIC COMPOSITE

A composite of ceramic and polymer has the best properties of both. It is light and flexible like the polymer and very strong in compression like the ceramic. However, it is not brittle. It could be used in construction as a cement and polymer composite and will last longer than either material alone. Again, this reduces co2 by lowering cement and oil use in production of polymers.

The elegance is found in the development of the material like bone over time with a sequence of events, using the same form to accomplish several functions. First the polymer is formed which gives off water to hydrate the cement, and the cement and polymer give off heat to set up the polymer. There is no outside energy needed for the reactions (as there usually is) and there is no effluent released (again as there usually is) because it is taken up as part of the second reaction forming cement. No one has ever developed such a material, and we hope this can find application in medicine as a bone replacement Bones develop over time from a sequence of events and I used this as the model.

What inspired this was an investigation into ceramic polymer composite systems and processing in which templating and bonding between the material phases yields improvements due to growth sequences based on bone formation.

The goal of this research was to develop composites with unique toughness and strength following the rules of bone growth by the templating of the ceramic through the polymer organization with intimate bonding between the two matrix materials. This occurs through careful growth sequences in the composite. The hollow structural fibers are placed first, and chemicals are released through the fibers to react with each other in a predetermined order and place. Templating occurs because one component reacts and forms first and the second reacts with it and follows its form. One reaction drives the other by taking up its effluent and giving off heat.

The emphasis is on the intimate connection between the phases for improved bonding due to careful growth sequences. It is known that epitaxial crystallization strategies, in which organized substrates are used to facilitate the formation and orientation of critical nuclei, have proven successful in mimicking the special properties of natural ceramics 1,2,3,4,5,6,7,8. In the process discussed here, first the polymer substrate polymerizes on which the ceramic then forms.

The focus of this research was: 1) templating and an intimate chemical bonding of a ceramic to a polymer so that the composite has no flaws at the interface and so has superior strength and toughness and 2) the fabrication method which creates this efficient composite in a prescribed way so that fibers and matrix interfaces relate in sequence over time. The processing steps and chemical composition in the composite already developed mimics natural ceramic/polymer composites.

This is a ceramic polymer composite which has templating and bonding between the polymer and the ceramic to produce unique chemical bonds without a coupling agent. It requires no mixing, no heating or other processing and produces no effluent or waste yet result in superior microstructure at the material interfaces.

The fabrication consists of hollow porous walled fibers or membranes, in woven configurations, being placed in a powder of a monomer and a ceramic. A polymer initiator flows through the fibers and is released along the length of the porous walled fibers into the powder matrix. The polymer initiator and polymer monomer powder react, forming a polymer, which gives off water that initiates the ceramic powder's solidification process. The removal of this polymer reaction product by the ceramic drives the polymer reaction to total completion; the chemicals are completely reacted with no waste products. Heat formed by the ceramic reaction then cures the polymer. The resultant composite has superior matrix interaction among the two matrix components as well as between the fibers and the matrix. Energy is saved in that no mixing or heating is required for the chemical reactions to occur. This efficient use of energy and chemicals produces a stronger and more durable, more economical composite. The production of a desired matrix product is limited by the equilibrium constant and the volume of chemical components originally present.

My physical system for this reaction is similar to a reaction separation tube, which uses catalytic reaction and extraction separation. A nylon polymer/cement combination was developed as the first system. The system used mechanical bonding. In the second composite made from a variation on the original chemistry, the bond between an acid nylon and the ceramic, the ceramic was bonded in a three-dimensional polymer scaffolding, which then appears to have influenced the form and crystallinity of the ceramic. This epitaxial growth was achieved with chemistry. The polymer acts as a template. The cement does the uptake of the polymer reaction by-product (water). Principles of equilibrium drive the first reaction, polymerization, towards completion.

This is a condensation polymerization reaction, which can be initiated without adding water, and thereby assuring that all the water to hydrate the cement comes from the condensation reaction. The heat of hydration of the cement sets the polymer; it is an exothermic reaction which drives the endothermic polymerization reaction. No heat addition nor mixing is required and no unused final product is present. Also, by varying the relative amounts of ceramic and polymer the composite can have more compressive or tensile strength. The closer chemical bond between the polymer and ceramic, the more stable the polymer and the attachment to the ceramic phase are, which ensures a less brittle composite and less degradation over time.

This polymer/ceramic composite will mimic nature by controlling the 1) chemical makeup and sequencing of fabrication, 2) the structure and form of the material. Shells or bone obtain their great toughness and strength because of careful control of these factors.

APPLYING RULES OF NATURAL CERAMICS TO COMPOSITES

Bone and other mineralized tissues are of interest as a model for composite materials because they are a composite of mineral and polymer. Improved polymer/ceramic composite materials can be created by applying the lessons learned from bone concerning composition, structure, adaptability, and self-repair. The chemical composition of bone, an organic-inorganic composite with intimate mechanical and chemical bonding at their interface, endows the material with a superior durability and toughness, which are desirable in a composite material. This composite should match bone in binding its polymer and ceramic phases through control of chemical sequencing and templating of the form. 2,8,10

The structure of the organic phase of bone is a three-dimensional network of collagen fibers with interpenetrating blood vessels (the inorganic phase forms upon and within the organic networks). This can be matched in a composite material by 3D weaving of organic fibers onto which the inorganic material forms. The adaptive properties of bone allow the material to react to environmental distress to most efficiently resist applied loads and prevent damage due to these outside factors by self-repair. 2,8,10

Controlling composition, structure, the process of formation, and its adaptive processes by using the rules of formation is the first step to mimicking bone to obtain superior properties. According to Arnold Caplan's writings about bone formation, "bone is made up of an oriented matrix which is secreted by bone-forming cells: the osteoblasts. This organic matrix is first made of structural molecules which serve as a scaffolding and which are laid down in a very precise, oriented pattern of fibrils into and onto which the inorganic crystalline phase forms. The formation of the first crystals of inorganic salt of calcium phosphate is referred to as the initiation or nucleation site which appear at regular intervals along this complex organic scaffolding of collagen laid down by osteoblast. Once nucleation has occurred, the next major process involves the continuation of crystalline growth from the nucleation sites outward along the fabric of the organic matrix and eventually between the molecules which serve as scaffolding. As crystal growth continues and forms a dense, inorganic matrix, there is a loss of organic components which are designed to reserve space in this matrix for the ever-expanding inorganic phase." 2,3

"The important landmarks of the organizational rules for bone growth that can be deduced are as follows:

1. Oriented multi-component organic matrix of fibrils as secreted by osteoblasts.
2. The formation of these oriented structural molecules serve as scaffolding (fibrils) and nucleation sites.
3. Initiation or nucleation of this inorganic, calcium phosphate crystalline phase on the sites.
4. The continuation of crystal growth with simultaneous rearrangement or elimination of components from the organic matrix."2,3

Adapting the rules of bone growth, according to Arnold Caplan [3] for these self-growing composite structures, includes making porous walled hollow polymer fibers that would release organic polymer chemicals into an inorganic matrix. The fibers would act as the organic template of fibrils onto which forms the strong structural bone-like composite from the matrix. The chemicals released from the fibers are designed to form a linked organic-inorganic matrix. The chemicals are monomers which release a chemical when polymerizing. The released chemical sets up the ceramic. The polymer tubes or fibers concentrate the polymer and bone-like inorganic substances on their surface. Ongoing self-healing over the life of the structure would be accomplished by reuse of the original void fibers or separate ones. These porous, walled, and hollow fibers would deliver repair chemicals when damage to the matrix occurs, such as cracking. In other research, we have shown this type of repair improves strength, toughness and ductility, therefore, there is no focus on it in this research activity.

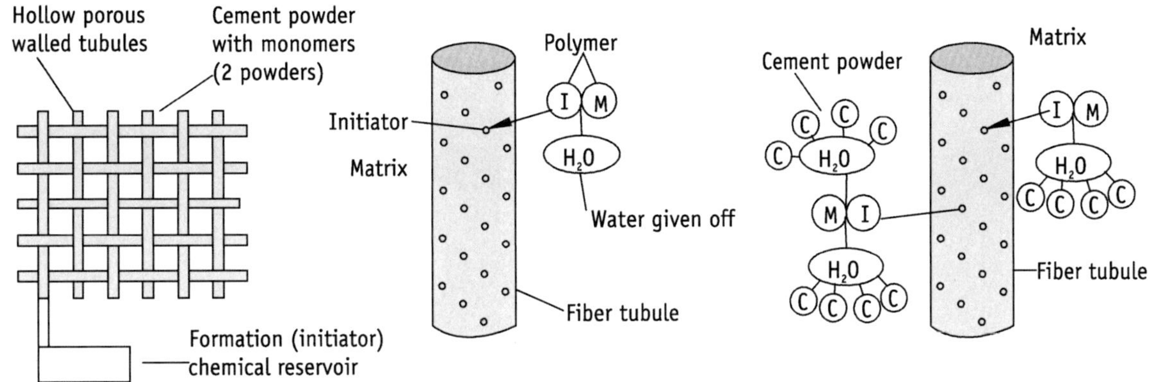

Diagram of the sequence of the processing to create the composite.

Development of the first system as a process for composite material production I have investigated nylon polymer reactions with cement and the method for inducing polymerization and ensuring the reaction of the ceramic. The chemistry was selected based on the following constraints: the reaction must take place near room temperature; the reactants and products must be relatively harmless; the resulting polymer must be reacted once it has been released in the matrix; the resulting polymer must not be water soluble; the polymer structure must provide a suitable interface for the link up with the ceramic; the chemical constituents must be widely available and relatively inexpensive; and the polymer must give off enough chemical to initiate reaction of the ceramic and not need mixing.

These reactions need to be symbiotic and self-driving. The polymerization produces the chemical product necessary to initiate and set up the ceramic, and the setup of the ceramic must absorb the byproduct of the polymerization reaction, thereby driving the polymerization reaction to produce more polymer (the DuPont Nylon 66 production reaction polymerization process usually required a vacuum to remove the water from the condensation reaction).

Time sequencing and development of a suitable fiber system for introducing the liquid monomer was essential in our experimentation. In the specific example of a condensation polymerization reaction hydrating cement, the polymerization produced the water necessary to hydrate the cement which absorbed the water byproduct of the condensation reaction, thereby driving the polymerization reaction to produce more nylon. The heat of the cement reaction contributed to the polymerization process of developing the polymer's strength. The delivery system evenly distributed the liquid monomer to all portions of the powder matrix. As this liquid reached the powder, the liquid monomer and the monomer in the matrix polymerized and gave off the water, which hydrated the cement. This is an exothermic reaction which emitted vapor. Covering the composite to prevent vapor loss captured these water vapors (which were necessary for hydration of the cement). The bonding appeared to be mainly mechanical with hints of templating.

X ACID NYLON AND CEMENT COMPOSITE

First, we developed a polymer cement combination in which there was some chemical bonding but mainly mechanical bonding. This composite material was weak in tension and compression but tougher than either nylon or cement alone.

A nylon polymer and cement were selected because of the variety of bonds and percentage of constituents possible for each individual application. Specifically, the condensation polymerization reaction to produce a mechanical bond was hexamethylene diamine and malic acid, which produces Nylon-6, 6. This selection was based initially on the fact that this polymerization could take place with little addition of heat, the reactants were inexpensive, they were less noxious than many other polymerization reactants, and that Nylon-6, 6 when polymerized formed a non-water-soluble polymer. Furthermore, this condensation reaction gave off the water necessary to hydrate the cement. The chemistry based on DuPont's method of nylon formation or nylon condensation polymerization is:

$$(n)HO\text{-}C(CH2)\ 4C\text{-}OH \quad + (n)H2N(CH2)\ 6\ NH2 \ =$$

$$HO\text{-}\ C(CH2)\ 4\ C\text{-}\ NH(CH2)\ 6\ NH\ \ n\ \text{-}H\ \ +(2n\text{-}1)H2O$$

Cement Hydration (assume water/cement ration of 0.5)

Samples were made to study the basic properties of the nylon composite in comparison with Nylon-6,6 and with cement. Butter stick samples (1"x1"x8") and compression cubes (2"x2"x2") were made of each material (Nylon-6,6, cement, and composite) to compare the bending and compressive strengths and behaviors of the materials see bending test table for results.

A set of samples was produced and tested to compare the compressive strength of the composite material to the strengths of nylon and cement control samples. The results above show that the composite is stronger in compression and fails in a manner consistent with a better modulus of elasticity and greater toughness than the sudden failures on the cement and nylon alone.

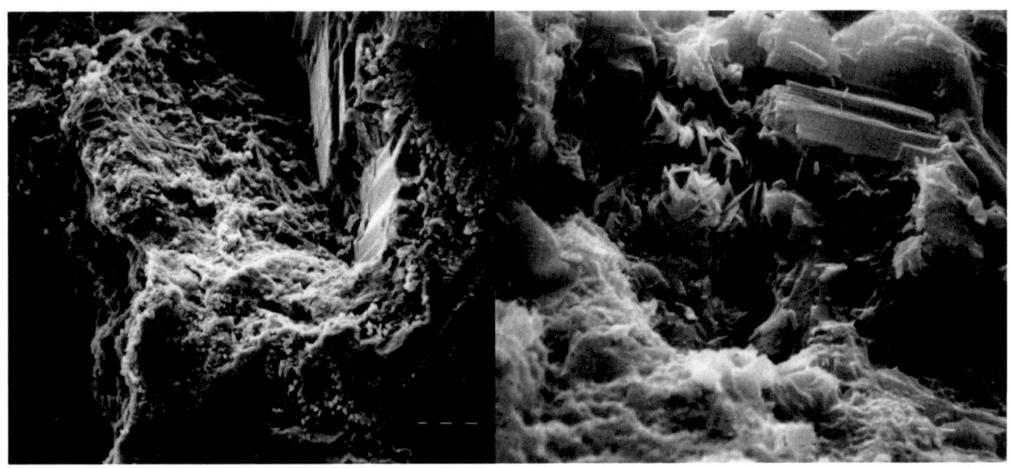

At left is a scanning electron microscope photo of malic acid nylon/ cement composite sample. Nylon is the block like material. At right is a SEM photo at one-tenth enlargement of photo at left.

In two scanning electron microscope photographs, at right in a 250X enlargement, it can be seen that the composite material is indeed a combination of the nylon and cement materials; therefore, the process of polymerization and cement hydration was successful. At 2500X enlargement, this is further confirmed, while it becomes apparent that the two materials seem to be primarily linked due to proximity in space. However, the resultant material at the points of contact seems to be notably different in appearance. This suggests that perhaps some chemical interaction between the constituents caused chemical bonding and templating.

The delivery systems were required to evenly distribute the liquid monomer to all portions of the powder matrix. Some of the ones tried are shown in the next figure. A series of straight tubules delivered chemical from the ends. Next a system was tried in which porous walled tubules delivered the chemicals and then a spiral formation of tubules was tried. Also, a series of porous walled membranes and fibers was designed. The spiral and porous walled tubules were most successful. As the liquid reaches the powder, the liquid monomer and the monomer in the matrix polymerizes and give off the chemical, which sets up the ceramic. The second nylon cement composite was made of adipic acid nylon/cement.

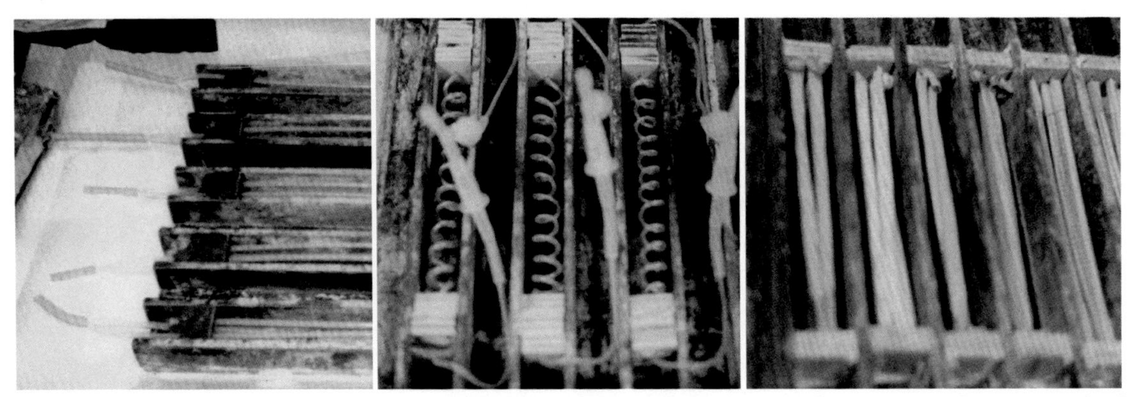

Photos showing the actual delivery systems used: Left, porous tubule fibers, Center, spiral porous tubule fibers Right, porous walled membrane delivery system.

Nylon was selected so that a variety of relationships with the ceramic were possible. These include a mechanical bond only or a 3D chemical bond in which the polymer heavily influences the form and crystallinity of the ceramic. A nylon polymerization reaction was sought that would produce a chemical bond to the cement, but with greater strength and more water production than that of malic acid nylon. The nylon polymer is the basis for doing this. By changing the acid used, one could achieve the chemical bond.

Various acids were considered that have a double bond such as succinic or malic acid. These formed the best samples and gave best resultant properties. See below for a diagram of this acid. The extra hydroxy group(s) offered the opportunity for chemical bonding. The enhanced chemical bonding of the polymer was then accomplished via this double bond. This combination was tested in compression and in bending. As seen, the strength was 3 times

$$\mathrm{HO(O{=})C{-}C(H){=}C(H){-}C({=}O)OH}$$

greater than with malic acid (but I was told later that we had used a liquid form of adipic acid which contained about three times the water to be given off by the polymerization reaction, therefore the extra cement hydrated could account for the nearly threefold rise in compressive strength). However, the increase in toughness and tensile strength is most likely due to the chemical bonding. Improved delivery system in the form of a helical single fiber helped ensure that all the chemicals reacted. Also, the adipic acid reaction was much hotter (at 170 degrees F) ensuring complete polymerization. This composite proved successful in terms of a more ductile failure consistent with better modulus of elasticity. There is the appearance in the SEMs of the cement taking on the form of the polymer as a template. These samples were tested after freezing in nitrogen to remove any necking of the polymer.

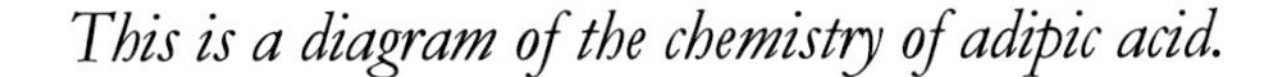

This is a diagram of the chemistry of adipic acid.

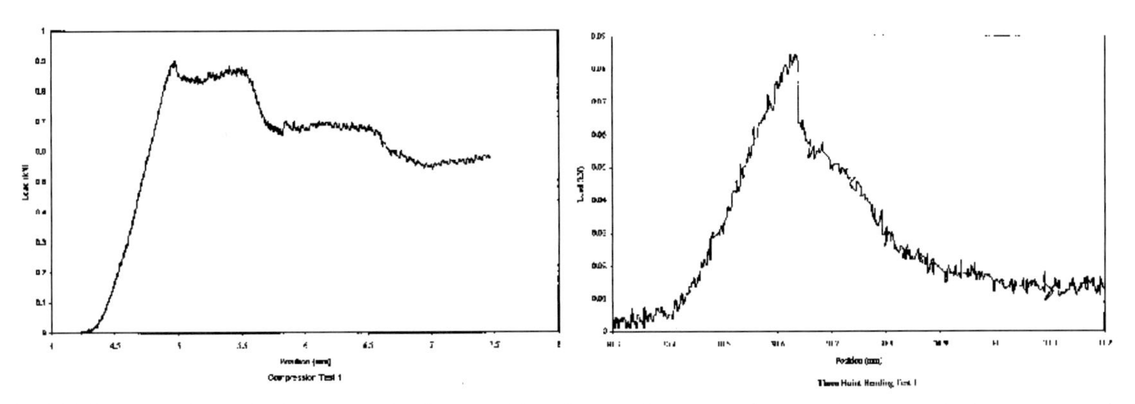

At left is typical load diagram of adipic acid nylon polymer and cement composite in compression. At right is typical load diagram of adipic acid nylon polymer and cement composite in bending.

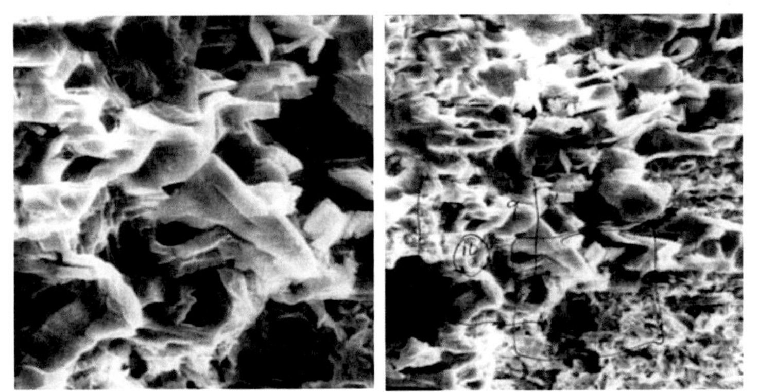

At left is a 3000x SEM photo of the frozen composite made with adipic acid and hexamethylene diamine and cement. The cement form takes on the shape of the polymer's stringy geometrical angles, using it as template. At right is a SEM photo of the composite at less enlargement.

SYSTEMS DESIGN AND RELATED REACTION SEPARATION

There are three aspects of the particular system as researched to date, namely: the epitaxial templating chemistries, the formal time sequencing process given by the release through fiber along the length of chemicals so one material crystallizes so the other can follow the pattern, and the synergistic reactions of the system so that one reaction drives the other and provides the heat for the other's reaction as well.

The system is a series of porous membranous fibers or tubules which release chemicals along their length so that the first reaction and byproduct are made and the reaction byproduct is removed along the length of the tube by the second reaction product. The formation of the first product, the polymer, acts as a template and generates an intimate bond for the ceramic, which is the second product. The uptake of the reaction product shifts the equilibrium so that the reactant use is increased as well as the production of both desired materials in the composite matrices. Most importantly, one desired product is produced first and can act as a scaffolding or template for the second desired material. Following the rules of bone growth inspired this whole process.

This specific system could be thought of as a generic reaction separation tube. True reaction separations usually are chemical systems used to improve the yields of the reaction, the production of desired products, and lower energy consumption. These are large industrial processes in which the reactor and separation apparatus are capital equipment. The proposed systems as invented uses a related idea but the "equipment" consists of a mold and fibers or tubules permanently located in the composite.

The fabrication and processing parameters including the dynamics of the reaction separation system includes initiator placement and compatibility of separation and reaction conditions for any certain system such as temperature, pressure. The impact of the initial formal structure of the 3-dimensional organic network of fibers is great on the chemical delivery. Specialized delivery systems allow for exact placement of chemicals with their reactants without mixing or heating. The next figure represents some of the various layouts of the fiber delivery morphologies. We have experimented with all formal structures of the 3-dimensional organic fiber network for chemical delivery except wound membrane.

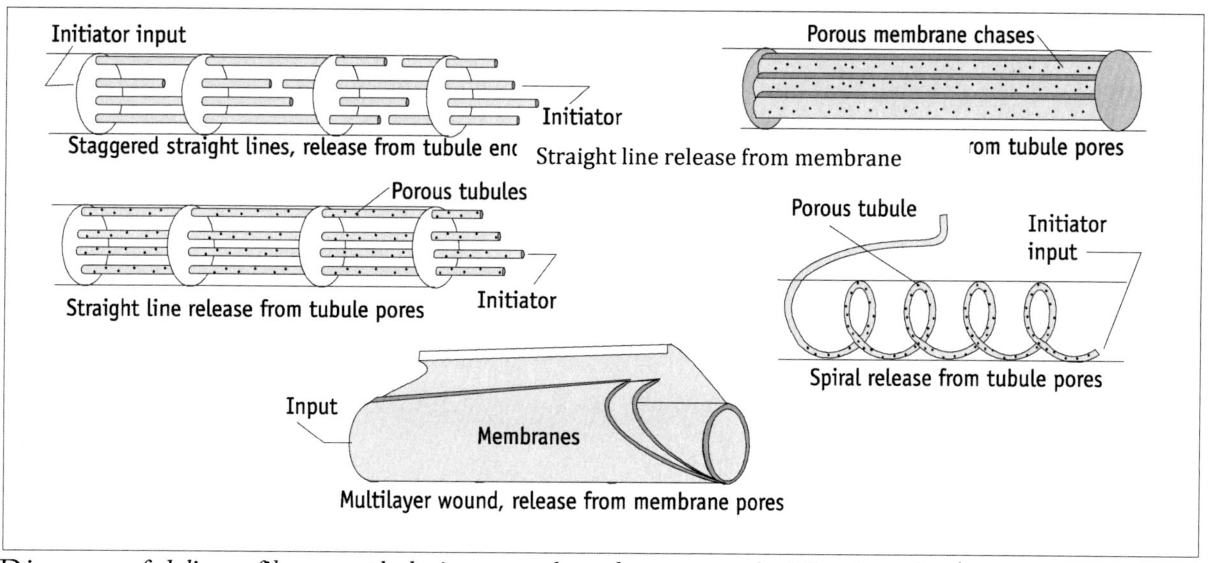

Diagrams of delivery fiber morphologies; some have been researched for the nylon/cement composite. 4-7

TEST CONCLUSIONS

The resultant properties and bond between phases yield toughness and strength and are superior to conventional ceramics alone due to the presence of organic bonding materials. The intimate connection between material phases is due to careful growth sequences; it is simulated by the process described in this work; by fibers, which are made first and then the matrix is grown around them.

The significance of this work is 1) that brittle ceramic materials will have ductile fibers bonded in the matrix and a polymer/ceramic inherently bonded and therefore the composite will have superior strength and flexure and toughness properties, 2) the morphology of the structure and controlled reaction between phases in the matrix will impart superior properties for greater durability of the composite

In applications, the advantages are spelled out:

1) As there is a polymer/ceramic bond, the composite is less prone to structural failure;
2) The formation at room temperature may cause a ceramic to form that has crystallinity approaching that of bone;
3) because there is both a polymer and a ceramic, the composite should be strong in compression, tension, and toughness;
4) the reaction allows a variety of mechanical and 3d chemical bonds as well as a variety of mix of proportions of the constituents; therefore the new material type can be

designed for each particular use;

5) the close chemical bond between the polymer and ceramic and the stability of the polymer and the attachment to the ceramic phase ensure a composite less prone to fracture failure over time and 6) the 3D scaffolding templated attachment has a good, strong bond

In future research, the approach will be to develop new candidate chemistry and delivery systems which we will then model, along with the two systems already developed to assess 1) the factors which influence the templating and bonding, 2) the influence of the delivery systems on the chemical reactions, 3) the influence of the delivery system on the chemical reaction system and in year, 4) the influence of the design for the delivery system on the reuse as a self-repair delivery system.

References cited

1. Peter C Rieke, Susan B. Bentjen, Barbara J Tarasevich, Thomas S Autry, and John Nelson "Synthetic Surfaces as Models for Biomineralization Substrates" *Mat. Res. Soc. Symp. Proc.* Vol. 174, 1990, pp. 69-80.

2. Brigid R Heywood and Stephen Mann, "Template-Directed Nucleation and Growth of Inorganic Materials", Adv. Mater. 1994, 6, No.1, pp. 9-19

3. Caplan, Arnold, "Bone Foundation, The Rules for Fabricating a Composite Ceramic". Mat. Res. Soc. Symp. Proc. Vol. 174, 1990, pp. 9-10

4. Mann, Stephen. Heywood, Brigid R. Rajam, Sundara. Walker, Justin B A. "Structural and stereochemical relationships between Langmuir monolayers and calcium carbonate nucleation." Journal of Physics D-Applied Physics. v 24 n 2 Feb 14 1991. p 154-164

5. Landau, E.M., R. Popovitz, et. al. "Langmuir monolayers designed for the oriented growth of glycine and sodium chloride crystals at air/water interfaces" *Mol. Cryst. Liq. Cryst.* 134, 1986. p323

6. Bianconi, P.A., B.J. Brisdon, S. Mann, et al "Polymeric-mediated crystallization of inorganic solids: calcite nucleation on poly(organosiloxane) surfaces" *Adv. Mater.* 5, 1993, p 49.

7. Mann, S., et. al. "Crystalization at inorganic-organic interfaces : biominerals and biomimetic synthesis" Science, 261, 1993,. p1286-1292.

8. Vincent, Julian F. V. *Structural Biomaterials*, Macmillan Press Ltd., London, 1982.

CONCLUSION

LOOKING FORWARD

I have described the paradigm of designing with natural processes and have shown the all-encompassing port project with examples of the design ideas in the other projects. All of the technologies are designed to take up the effluents of a fossil fuel economy in order to make a better world for tomorrow. All but the port project were made at full scale and usually with a commercial company. From concept of the paradigm to full scale, the projects shown have practical applications of designing in a way that works with and uses the processes of nature. I hope to inspire interest in this kind of environmental preservation and subsequently innovative research.

References written by Carolyn Dry, Ph.D.

CHAPTER 1:

GENERAL CONCEPTS AND PROJECT OVERVIEWS

No references for this section.

CHAPTER 2:

MAKING OCEAN PORTS USING THE CHEMISTRY OF SEAWATER

"Approaches to Ocean Resource Development", Sponsored by the U.S. Office of Naval Research through Texas A and M University, August 31, 1975

CHAPTER 3:

FUNCTIONS OF THE PARADIGM: ASH BUILDING PRODUCTS AND USE OF WASTE MATERIALS

"The Potential Use of Waste (Ash) Material for EIFS Insulation and Related Components," Development, Use, and Performance of Exterior Insulation and Finish Systems (EIFS), ASTM STP 1187, M. F. Williams, R. G. Lampo, and R. G. Reitter II, eds., pp. 351-358, American Society for Testing and Materials, Philadelphia, PA, 1995

"Changing Exterior Applied Insulation, Wall Sheathing, and Panels Made From Waste Materials," Journal of the Canadian Ceramic Society, Vol. 63, no. 1, pp. 54-58, Feb. 1994

"Sintered Fly Ash and Acid Composition Materials for Three Uses: Structural Sandwich Panels, Exterior Applied Insulation, and Insulation Backerboards," Materials & Structures, 1995.

"Microstructure Optimization of Sintered Fly Ash/Acid Composition for Use as Structural Thermal Insulation in Buildings," Journal of Thermal Insulation and Building Envelopes, Vol. 19, April 1996, pp. 336-347

"Low Cost Exterior Building Insulation Panels from Sintered Fly Ash have Improved Properties and Reduce Environmental Pollution," Material Technology, Nov/Dec 1996, pp. 228-229

"Ceramic Structural Sandwich Panels Made from Sintered Fly Ash, Bottom Ash, Phosphoric Acid," 1st International Symposium on the Science of Engineering Ceramics, Memorial Symposium of the 100th Anniversary of the Ceramics Society of Japan, edited by S. Kunura and K. Niihara, October 21-23, 1991, Koda, Japan, pp. 595-600

"Characterization of two sintered fly ash/acid composites." Canadian Ceramics Quarterly/Journal of the Canadian Ceramic Society, 63(1), 54-58. 1994.

"Proceedings of the 3rd rilem international symposium on autoclaved aerated concrete." 14-16 October 1992. Advances in autoclaved aerated concrete. Editor Folker H. Wittmann. p. 315-320)

"Durability of Sandwich Panels Made from Bonded/Sintered Fly Ash and Bottom Ash," Proceedings of the 9th International Ash Use Symposium, Electric Power Research Institute, January 1991, pp. 77—1 to 77—21.

"Durability of Sandwich Panels Made from Bonded/Sintered Fly Ash and Bottom Ash," Proceedings of the 1st International Symposium on the Science of Engineering Ceramics, Koda—cho, Japan, October 21 - 23, 1991.

"Sintered Coal Ash / Flux Materials For Building Materials." Materials and Structures. With Bukowski, J., Meier, J., 30 Oct., 2002).

"Use of Recycled Materials (Ash) to Address the Performance Needs for Durability of Prefabricated Building Materials, Blocks, and Panels," Proceedings CIB `92 World Building Congress, Poster Volume, May 18-22, 1992, Montreal, Quebec, Canada Institute for Research in Construction.

"Sintered Foamed Fly Ash and Acid Composition and Fly Ash Floater and Acid Material as Alternatives to Cellular Concrete," Advances in Autoclaved Aerated Concrete, edited by F. H. Wittman, Zurich, Switzerland, October 14-16, 1992, A. A. Balkema Publishers, Rotterdam, Netherlands, pp. 315-320

"Novel Materials for Insulation," Construction Engineering Research Laboratory (CERL), U.S. Army, 8 pages, 1993 National Research Council of Canada, p. 95. With Kathy Best.

CHAPTER 4:

SELF-REPAIRING CONCRETE

"Building Materials, Which Evolve and Adapt Over Time--Use of Encapsulation Technology and Delayed Reactions." University of Illinois, Proceedings of the 1st International Architectural Research Centers Consortium Conference/ARCC '88, Urbana-Champaign, Illinois, November 13-15, 1988, pp. 6-11

"Building Materials Which Evolve and Adapt Over Time; Use of Encapsulation Technology and Delayed Reactions." Proceedings of the Conference of International Council for Building Research, Studies and Documentation (CIB), Paris, France, June 19-23, 1989, pp. 391-399

"Technology Adoption--Survey of Government Agencies, Facilities, Industry Facilities, and Vendors," Construction Engineering Research Laboratory (CERL), U.S. Army, 36 pages, 1989

"Alteration of Matrix Permeability, Pore and Crack Structure by the Time Release of Internal Chemicals." Proceedings ACS/NIST Conference on Advances in Cementitious Materials, edited by S. Mindess, American Ceramic Society, Inc., co-sponsored by National Institute of Standards and Technology, Gaithersburg, Maryland. Meeting held July 22-26, 1990, Elsevier Publisher, pp. 729-768

"Design of Systems for Time-delayed activated Internal Release of Chemicals in Concrete from Porous Aggreagtes and Prills", Virginia Polytechnic Institute and University, PhD Dissertation in Environmental Design and Planning, January 1991

"Passive Tuneable Fibers and Matrices." Proceedings of the International Conference on Electrorheological Fluids, edited by R. Tao, Southern Illinois University, October 14-15, 1991, Carbondale, Illinois, World Scientific Publishing, Singapore, pp. 494-498

"Timed Release of Chemicals into Hardened Matrices Cementitious for Crack Repair, Rebonding Fibers, and Increasing Flexural Toughening," Fracture Mechanics, 25th Volume, ASTM STP 1220, F. Erdrogan, editor, pp. 268-282, American Society for Testing and Materials, Philadelphia, PA, 1995

"Monitoring and Repair by Release of Chemicals in Response to Damage," Intelligent Civil Engineering Materials & Structure, an ASCE special publication, 1997 (in press)

"Liquid Core Optical Fibers for Crack Detection and Repairs in Concrete Matrices," Special Technical Publication of Workshop on Fiber Optics, Newark, NJ

"Passive Smart Materials for Sensing and Actuation," Journal of Intelligent Material Systems and Structures, Vol. 4, no. 3, pp.415-418, Blacksburg, Virginia, July 1993

"Matrix Cracking Repair and Filling Using Active and Passive Modes for Smart Timed Release of Chemicals From Fibers Into Matrices," Journal of Smart Materials and Structure, Vol. 3, no. 2, pp. 118-123, June 1994

"Smart Fibers that Sense and Repair Damage in Concrete Materials," Materials Technology, Vol. 11, no. 2, March/April 1996, pp. 52-54

"Building Materials That Self-Repair," Architectural Science Review, Vol. 40, Melbourne, Australia, June 1997, pp. 45-48

"Design of Inexpensive Self-growing, Self-repairing Building Construction Materials Which Perhaps Improve the Environment," special issue of the Electronic Green Journal, Fall 1998

"Smart Building Materials Which Prevent Damage or Repair Themselves." Material Research Society Symposium Proceedings on Smart Materials Fabrication & Materials for Micro-Electrical-Mechanical Systems, Vol. 276, editors A. P. Jardine, et al., San Francisco, California, April 28-30, 1992, Materials Research Society Publishers, Pittsburg, Pennsylvania, pp. 311-314

"Smart Materials Which Sense, Activate and Repair Damage; Hollow Porous Fibers in Composites Release Chemicals fromFibers for Self-Healing Damage Prevention, and/or Dynamic Control," Proceedings on First European Conference on Smart Structures and Materials, edited by B. Culshaw, et al., Glasgow, Scotland, May 12-14, 1992, SPIE Volume 1777, Institute of Physics Publishing, Bristol, England, and EOS/SPIE and IOP, EUROPTO Series Publishing Ltd., pp. 367-371.

"Pavements Which are Self-Healing by the Release of Repair Chemicals Upon Demand," Smart Pavement Conference, Dallas, Texas, December 5, 1995, pp. 52-56

"Smart Materials that Self-Repair by Timed Release of Chemicals," SPIE 1996 Symposium, Smart Materials, Structures & MEMS, Banglore, India, December 11-14, 1996 (in press)

"Release of Smart Repair Chemicals for the In-Service Repair of Bridges and Roadways," SPIE 1996 Symposium, Smart Materials, Structures & MEMS, Banglore, India, December 11-14, 1996 (in press)

"Repair and prevention of damage due to transverse shrinkage cracks in bridge decks", published in SPIE Proceedings Volume 3671: Smart Structures and Materials 1999: Smart Systems for Bridges, Structures, and Highways, May 1999

"Improvement in Reinforcing Bond Strength in Reinforced Concrete with Self-Repairing Chemical Adhesives," Smart Systems for Bridges, Structures, and Highways, Proceedings of SPIE's Smart Structures and Materials Conference, 1997, pp. 44-50

"Smart Bridge and Building Materials in Which Cyclic Motion is Controlled by Internally Released Adhesives," Smart Systems for Bridges, Highways, and Structures, Symposium on Smart Materials and Structures '96, Proceedings of SPIE's Smart Structures and Materials Conference, San Diego, CA, February 1996

"Preserving Performance of Concrete Members Under Seismic Loading Conditions," Smart Systems for Bridges, Structures, and Highways, Proceedings of SPIE's Smart Structures and Materials Conference, San Diego, March 1-5, 1998. With J. Unzicker,

"Self repair of impacts, higher energy impacts, and earthquake damage in critical targets such as infrastructure components made of polymers and concrete", published in SPIE Proceedings Volume 6531: Nondestructive Characterization for Composite Materials, Aerospace Engineering, Civil Infrastructure, and Homeland Security 2007, April 2007

"Use of embedded self-repair adhesives in certain areas of concrete bridge members to prevent failure from severe dynamic loading" with Jacob Unzicker, Published in Proceedings Volume 3675: Smart Structures and Materials 1999: Smart Materials Technologies. July 1999
"Preserving performance of concrete members under seismic loading conditions"
And Jacob Unzicker, Univ. of Illinois/Urbana-Champaign (United States)
Published in SPIE Proceedings Volume 3325: Smart Structures and Materials 1998: Smart Systems for Bridges, Structures, and Highways June 1998
"A Comparison Between Adhesive and Steel Reinforced Concrete in Bending," Journal of Cement and Concrete,
"Structural Control During and After Seismic Events by Timed Release of Chemicals for Damage Repair in Composites Made of Concrete or Polymers, First World Conference on Structural Control, Proceedings, Volume 2, edited by G. W. Housner, S. F. Masri, A. G. Orassia, International Assoc. for Structural Control, Los Angeles, California, August 3-5, 1994, pp. TAI-60 - TAI-65
"A Time Release Technique for Corrosion Prevention," Journal of Cement and Concrete, with Melinda Corsaw, Summer 1999.
"Smart Concrete," Progressive Architecture, July 1995, pp. 92, 93
"Three Part Methylmethacrylate Adhesion System as an Internal Delivery System for Smart Responsive Concretes," Smart Materials and Structures, 1996, pp. 297-300
"Three Designs for the Release of Internal Sealants, Adhesive, Waterproofing Chemicals into Concrete to Reduce Permeability and Crack Damage, Journal of Cement and Concrete Research, 30,2000, pp 1969=1977
"Passive Tuneable Fibers and Matrices," International Journal of Modern Physics B, Vol. 6, Nos. 15 & 16, pp. 2763-2771, World Scientific Publishing Co., Rivers Edge, New Jersey, 1992

CHAPTER 5:

SELF-REPAIRING POLYMERS

"Smart Multiphase Composite Materials Which Repair Themselves by a Release of Liquids Which Become Solids," in Smart Structures and Materials 1994: Smart Materials, Proceedings SPIE 2189, V. K. Varadan, Editor, 1994, pp. 62-70
"Passive Smart Self-Repair in Polymer Matrix Composite Materials," in Smart Structures and Materials 1993: Smart Materials, Proceedings, SPIE 1916, V. K. Varadan, Editor, 1993, pp. 438-444. With N. Sottos.
"Release of Repair Chemicals from Fibers In Response to Damage in Polymer Composites, Proceedings of First International Conference on Composites Engineering, edited by D. Hui, New Orleans, Louisiana, August 28-31, 1994, pp. 1061-1062
"Procedures Developed for Self-Repair of Polymer Matrix Composite Materials," Composite Structures, 35, Paisley, UK, 1996, pp. 263-269
"Thixotropic action of self-repairing chemicals to increase strength at first impact", published in SPIE Proceedings Volume 9059: Industrial and Commercial Applications of Smart Structures Technologies 2014, May 2014
"Comparison of self repair in various composite matrix materials", published in SPIE Proceedings Volume 9059: Industrial and Commercial Applications of Smart Structures Technologies 2014, May 2014
"A progression of damage repair capability in self-repairing composites" published in SPIEProceedings Volume 9059: Industrial and Commercial Applications of Smart Structures Technologies 2014, May 2014
"Self-repairing composite walls for pressurized space habitats", published in SPIEProceedings Volume 9801: Industrial and Commercial Applications of Smart Structures Technologies 2016, September 2016
"Testing of self-repairing composite airplane components by use of CAI and the release of the repair chemicals from carefully inserted small tubes" published in SPIE Proceedings Volume 6527: Industrial and Commercial Applications of Smart Structures Technologies 2007, April 2007
"Self-repairing composites for airplane components" published in SPIE Proceedings Volume 6932: Sensors and Smart Structures Technologies for Civil, Mechanical, and Aerospace Systems 2008, May 2008
"Self repairing composites for drone air vehicles" published in SPIE Proceedings Volume 9433: Industrial and Commercial Applications of Smart Structures Technologies 2015, May 2015

CHAPTER 6:

SELF-SENSING IN CONCRETE AND POLYMER COMPOSITES

"A Novel Method to Detect Crack Location and Volume in Opaque and Semi-Opaque Brittle Materials," Journal of Smart Material and Structures, Vol. 6, 1997, pp. 35-39
"X-ray and Fiber Optic Detection of Cracks in Concrete Using Adhesive Released from Glass Fibers," Smart Sensing, Processing, and Instrumentation, Proceedings of SPIE's Smart Structures and Materials Conference, 1997
"Crack and Damage Assessment in Concrete and Polymer Matrices Using Liquids Released Internally from Hollow Optical Fibers," Smart Sense Processing, Symposium on Smart Materials and Structures '96, Proceedings of SPIE's Smart Structures and Materials Conference, San Diego, CA, February 1996, pp. 448-451

"Smart Materials for Sensing and/or Remedial Action to Reduce Damage to Materials." Proceedings of ADPA/AIAA/ASME/SPIE Conference on Active Materials and Adaptive Structures, edited by Gareth Knowles, November 4-8, Alexandria, Virginia, 1991, Institute of Physics, Publishing, London, Great Britian, pp. 191-194
"Passive Smart Materials for Sensing and Actuation." Proceedings: Conference on Recent Advances in Adaptive and Sensory Materials and Their Applications, edited by C. A. Rogers and R. C. Rogers, Virginia Polytechnic Institute and State University, Blacksburg, Virginia, April 27-29, 1992, Technomic Publishers, Lancaster, England, pp. 207-223
"Adhesive Liquid Core Optical Fibers for Crack Detection and Repairs in Polymer and Concrete Matrices," in Smart Structures and Materials 1995: Smart Sensing, Processing and Instrumentation, Proceedings SPIE 2444, W. B. Spillman, Jr., Editor, 1995, pp. 410-413
"Crack and damage assessment in concrete and polymer matrices using liquids released internally from hollow optical fibers"with William McMillan, published in Proceedings Volume 2718: Smart Structures and Materials 1996: Smart Sensing, Processing, and Instrumentation, May 1996

"Self sensing composites with emi shielding and self repair", published in SPIE Proceedings Volume 9435: Sensors and Smart Structures Technologies for Civil, Mechanical, and Aerospace Systems 2015, May 2015

CHAPTER 7:

RECYCLING

No references for this section.

CHAPTER 8:

CO_2 ABSORBING COATINGS

No references for this section.

CHAPTER 9:

SELF-FORMING POLYMER/CERAMIC COMPOSITE

"A Biomimetic Bone-like Polymer Ceramic Composite: With Emphasis on the Initial Structural Form and Composition and Ability to Adapt During Fabrication and Throughout the Material's Life," Smart Systems for Bridges, Structures, and Highways, Proceedings of SPIE's Smart Structures and Materials Conference, San Diego, March 1-5, 1998
"Polymer Ceramic Composite which Mimics Bone Formation," SPIE Symposium on Smart Structures and Materials, conference on Electro-active Polymer Actuators and Devices, Newport Beach, California, March 1-5, 1999
"Self-repairing, self-forming, and self-sensing systems for ceramic/polymer composites", published in SPIE Proceedings Volume 4701: Smart Structures and Materials 2002: Smart Structures and Integrated Systems, July 2002
"Exploration of electric properties of bone compared to cement: streaming potential and piezoelectirc properties", published in SPIE Proceedings Volume 9429: Bioinspiration, Biomimetics, and Bioreplication 2015, May 2015
"Paradigm for design of biomimetic adaptive structures." Published in SPIE Proceedings Volume 9429: Bioinspiration, Biomimetics, and Bioreplication 2015. May 2015
"Passive self repairing and active self sensing in multifunctional polymer composites." Published in SPIE Proceedings Volume 6928: Active and Passive Smart Structures and Integrated Systems 2008. May 2008
"Sensing of repair in chemically self-repairing composites." Published in SPIE Proceedings Volume 9059: Industrial and Commercial Applications of Smart Structures Technologies 2014. May 2014
"Factors affecting self-repairing of composites." Published in SPIE Proceedings Volume 4935: Smart Structures, Devices, and Systems. November 2002
"Smart materials which sense, activate and repair damage; hollow porous fibers in composites release chemicals from fibers for self-healing, damage prevention, and/or dynamic control." Published in SPIE Proceedings Volume 1777: First European

Conference on Smart Structures and Materials. May 1992
"Design of self-growing, self-sensing, and self-repairing materials for engineering applications." Published in SPIE Proceedings Volume 4234: Smart Materials. April 2001
"Release of smart chemicals for the in-service repair of bridges and roadways." Published in SPIE Proceedings Volume 3321: 1996 Symposium on Smart Materials, Structures, and MEMS. April 1998
"Biomimetic rules for design of complex adaptive structures." Published in SPIE Proceedings Volume 4512: Complex Adaptive Structures. October 2001
"Self-forming polymer ceramic composite made by an in-situ process to yield superior microstructrue while using materials and energy efficiently." Published in SPIE Proceedings Volume 4234: Smart Materials. April 2001
"Passive smart self-repair in polymer matrix composite materials." With Nancy R. Sottos, Univ. of Illinois/Urbana-Champaign (United States). Published in SPIE Proceedings Volume 1916: Smart Structures and Materials 1993: Smart Materials. July 1993
"Smart multiphase composite materials that repair themselves by a release of liquids that become solids." Published in SPIE Proceedings Volume 2189: Smart Structures and Materials 1994: Smart Materials. May 1994
"Polymer-ceramic composite that mimics bone formation." Published in SPIE Proceedings Volume 3669: Smart Structures and Materials 1999: Electroactive Polymer Actuators and Devices. May 1999
"Damage assessment using liquid-filled fiber optic systems." Published in SPIE Proceedings Volume 3321: 1996 Symposium on Smart Materials, Structures, and MEMS. April 1998
"Self-repairing of composites." Published in SPIE Proceedings Volume 5055: Smart Structures and Materials 2003: Smart Electronics, MEMS, BioMEMS, and Nanotechnology. July 2003
"X-ray and fiber optic detection of cracks in concrete using adhesive released from hollow glass fibers." Published in SPIE Proceedings Volume 3042: Smart Structures and Materials 1997: Smart Sensing, Processing, and Instrumentation. June 1997
"Adhesive liquid core optical fibers for crack detection and repairs in polymer and concrete matrices." Published in SPIE Proceedings Volume 2444: Smart Structures and Materials 1995: Smart Sensing, Processing, and Instrumentation. April 1995
"Biomimetic bonelike polymer cementitious composite." with Carrie Warner. Published in SPIE Proceedings Volume 3040: Smart Structures and Materials 1997: Smart Materials Technologies. February 1997
"Smart earthquake-resistant materials: using time-released adhesives for damping, stiffening, and deflection control." Published in SPIE Proceedings Volume 2779: 3rd International Conference on Intelligent Materials and 3rd European Conference on Smart Structures and Materials. April 1996
"Smart bridge and building materials in which cyclic motion is controlled by internally released adhesives." Published in SPIE Proceedings Volume 2719: Smart Structures and Materials 1996: Smart Systems for Bridges, Structures, and Highways April 1996
"Current research in timed release of repair chemicals from fibers into matrices." Published in SPIE Proceedings Volume 2361: Second European Conference on Smart Structures and Materials. September 1994
"Polymer ceramic composite that follows the rules of bone growth." with Carrie Warner, Published in SPIE Proceedings Volume 3324: Smart Structures and Materials 1998: Smart Materials Technologies. July 1998
"Self-repair of cracks in brittle material systems." Published in SPIE Proceedings Volume 9800: Behavior and Mechanics of Multifunctional Materials and Composites 2016. June 2016
"Two intelligent materials, both of which are self-forming and self-repairing; one also self-senses and recycles." Published in SPIE Proceedings Volume 2779: 3rd International Conference on Intelligent Materials and 3rd European Conference on Smart Structures and Materials. April 1996
"Improvement in reinforcing bond strength in reinforced concrete with self-repairing chemical adhesives." Published in SPIE Proceedings Volume 3043: Smart Structures and Materials 1997: Smart Systems for Bridges, Structures, and Highways. May 199

Made in the USA
Lexington, KY
15 September 2018